A Psychohistory of Metaphors

A Psychohistory of Metaphors

Envisioning Time, Space, and Self throughout the Centuries

Brian J. McVeigh

LEXINGTON BOOKS
Lanham • Boulder • New York • London

Published by Lexington Books
An imprint of The Rowman & Littlefield Publishing Group, Inc.
4501 Forbes Boulevard, Suite 200, Lanham, Maryland 20706
www.rowman.com

Unit A, Whitacre Mews, 26-34 Stannary Street, London SE11 4AB

British Library Cataloguing in Publication Information Available

Library of Congress Cataloging-in-Publication Data

Names: McVeigh, Brian J., author.
Title: A psychohistory of metaphors : envisioning time, space, and self throughout the centuries / Brian J. McVeigh.
Description: Lanham : Lexington Books, [2016] | Includes bibliographical references and index.
Identifiers: LCCN 2015051388 (print) | LCCN 2016002303 (ebook) | ISBN 9781498520287 (cloth : alk. paper) | ISBN 9781498520294 (electronic)
Subjects: LCSH: Space and time–Psychological aspects. | Social history–Psychological aspects. | Metaphor–Psychological aspects. | Metaphor–Philosophy.
Classification: LCC BF467 .M4885 2016 (print) | LCC BF467 (ebook) | DDC 115–dc23
LC record available at http://lccn.loc.gov/2015051388

∞™ The paper used in this publication meets the minimum requirements of American National Standard for Information Sciences Permanence of Paper for Printed Library Materials, ANSI/NISO Z39.48-1992.

Printed in the United States of America

Contents

Foreword vii

Acknowledgments and Notes to Reader ix

Prologue: Explaining History's "Inward Turn" 1

1 My Search for Heaven and Hell 3

2 Purposes and Premises: Tracing the Trajectories of Human Experience 9

3 The Magic of Metaphors: How Our Minds Make the World 19

4 Unpacking the "Black Box" of Conscious Interiority 31

Part I: Space: Hollowing Out the Person 39

5 Envisioning the Invisible: Spatializing the Soul 41

6 Invoking Introspectable Worlds 49

7 The Collapse of Premodern Cosmology 59

Part II: Psyche: The Origins of Scientific Psychology 69

8 The Foundations of the Modern Study of Mind 71

9 The Great Cosmic Split: Dualism 77

10 Reactions to the Cartesian Split 87

11 Early Psychology: Making Visible the Contents of the Soul 97

Part III: Time: Modern Millenarianism and Politics as "Progress" 107

12 Meta-Framing Time: The Invention of History 109

13 Liberating the Psyche: The Emerging Faith in Progress 117

14 The History of Humankind: Climbing the Ladder of Civilization 127
15 Envisaging the Future as Paradise 135

Part IV: Self: Turning the World Inside Out 141
16 The Changeable Self through the Centuries 143
17 The Narratized Individual as Social Actor 155
18 The Self as Mirror in Historical Perspective 161
19 The Birth of Modern Psychology 169

Epilogue: Visualizing New Vistas of Modernity and Selfhood 175
20 The Therapeutic Turn 177
21 Self-Idolatry: The Dark Side of the Psychotherapeutic Society 183
22 Modern Spatiality, the Soul, and the Psyche 193

Appendix A: How to Open the "Black Box": Cultural Psychology 197

Appendix B: What Conscious Interiority Is Not 201

Appendix C: Spaces: Real and Imaginary 209

Appendix D: The Visible, Invisible, and Introspectable 211

References 213

Index 225

About the Author 233

Foreword

When asked about their thoughts on consciousness, most people view it as something that is biological, innate in all people, and present to a lesser degree in most higher mammals. This misconception arose largely as an extrapolation from Darwin's theory of evolution: If all life evolved, then consciousness must also be a by-product of evolution. One of the reasons for its persistence was the fact that, from roughly the 1920s to the 1960s, American psychology was dominated by behaviorism. Behaviorism, started by John B. Watson and later expanded upon and popularized by B. F. Skinner, emphasized the study of observable behavior and ignored (or even denied) inner mental life. Behaviorism resonated with the movement to make psychology as a field of study "more scientific" (and less philosophical) and the American ideal of unlimited potential—a form of rebellion against ideas about heredity. Consciousness, although previously a popular topic, became taboo.

While the 1960s saw the rise of cognitive psychology and the decline of behaviorism, the study of consciousness was still largely ignored. Inner mental life was widely viewed as a topic that would not yield productive research, and in academia an atmosphere of political correctness strongly discouraged the study of cross-cultural or even historical mental differences.

It was in this intellectual climate that, in early 1977, the psychologist Julian Jaynes released his seminal book, *The Origin of Consciousness in the Breakdown of the Bicameral Mind.* In it he proposed that consciousness—by which he means our ability to introspect and not mere sense perception, which we share with all animals—was not innate but a learned process based on metaphorical language, and taught to each successive generation. According to Jaynes, human psychology underwent a profound transformation as recently as three thousand years ago.

Metaphor is far more pervasive throughout language than most realize, and Jaynes argues that metaphor is not just linguistic embellishment but that it facilitates thought processes that would not otherwise be possible. All of our terms for mental abstractions have their basis in metaphor, and metaphors are instrumental in our ability to introspect—our inner life or internal "mind-space" being built up through metaphors of physical space.

The advent of inner mental life brought about profound changes in human culture. The metaphor of space allowed us to think about the concept of time—indeed, it is only through the metaphor of space that we can think about time. We could, for the first time, think about our lives on a continuum and contemplate our own death. Human emotions changed as for the first time we could introspect and ruminate on past events. Consciousness operating on our emotions allows shame to be transformed into guilt, fear into anxiety, and anger into hatred.

Despite Jaynes's profound insights and a growing body of literature on the important role of metaphor in language, little progress has been made in the intervening decades toward understanding the mechanisms by which metaphorical language creates an internal mind-space. Brian McVeigh's latest book, *A Psychohistory of Metaphors*, is thus a long overdue and welcome contribution to the subject.

As a graduate student in anthropology at Princeton University, McVeigh was a student and friend of Julian Jaynes, and the influence of Jaynes's ideas can be seen in the present volume. In a world of ever-increasing scientific specialization, McVeigh, like Jaynes, brings a refreshingly interdisciplinary approach, blending his interests in anthropology, psychology, and ancient history with decades of cross-cultural research (he spent many years teaching in Japan and has written extensively on Japanese culture).

Brian McVeigh extends Jaynes's ideas on metaphor and thought; expands upon the different features of consciousness; explains the interrelatedness of our conceptions of time, space, and the self; and explores some of the implications of our newly learned inner life—the consequences of our consciousness. His ideas constitute a significant step forward to understanding both the metaphorical basis of thought and the human condition.

—Marcel Kuijsten,
Executive Director of the Julian Jaynes Society

Acknowledgments and Notes to Reader

It is hard to find, I imagine, graduate students who did not have to suffer during their formative years some amount of arrogance, whether from individuals or institutions. What a welcome relief, then, to have spent time with Julian Jaynes while I was a graduate student at Princeton University. His great insights placed him in awe before the complexity and subtlety of what we call psyche, and his humility and modesty were the ultimate muse for this book. For this I will always be grateful.

I would also like to thank Marcel Kuijsten, Executive Director of the Julian Jaynes Society. His encouragement of and commitment to the world of great ideas is greatly appreciated. I want to also express my gratitude to an anonymous reviewer who helped me strengthen my arguments. As always, my father, mother, Bill, Kristin, and Laurel, deserve my thanks for always being there. Special thanks go to my father-in-law and late mother-in-law, who have provided me with a restful abode during the summer months ideal for thinking and writing. And finally, my wife, Lana, deserves my deepest gratitude.

Throughout this book I distinguish between the academic discipline of *Psychology* (capitalized “P”) and the *psychological* (lowercase “p”) or what since the late nineteenth century has described processes defined as mind, emotion, perception, and cognition.

Prologue

Explaining History's "Inward Turn"

Chapter One

My Search for Heaven and Hell

When I was a young student in Catholic school, I asked my teachers where, exactly, do Heaven and Hell exist. To a young mind concerned about the possible realities of Hell, the nuns and priests never seemed quite able to satisfactorily answer my question. They would vaguely respond that the next world was not just a place but a sort of state of spiritual grace. I remember once being told that Heaven existed within me. This answer left me even more confused. Eventually I concluded that Heaven and Hell were semi-spatial, partially somewhere and partially nowhere, somehow existing both in the tangible cosmos and in my soul, or as St. Augustine wrote about the location of memories, they are in "an inner place, which is yet no place."[1]

The location—or nonlocation—of Heaven and Hell invokes issues about the relation between two types of space: physical space and imaginary space. People, of course, have been metaphorically employing spatial relations for centuries to depict difficult-to-describe things or situations. Hierarchical social relations are a good example; i.e., almost all cultures rely on tropes of "above" or "below" to characterize, well, who is "on top" and who is "on the bottom." What we often call "space," then, is not necessarily the three-dimensional variety of physical reality. Placeness may be unreal or imagined, e.g., mind-space, cyberspace, or a communal "space" of like-minded individuals who are not necessarily physically gathered together.

However, because we are so familiar with applying spatial tropes in daily life, a tendency to neglect some deeper truths has taken root. Consider how imaginary space, so significant for how we experience the world, is also crucial for how we conceive time. Indeed, according to Julian Jaynes, "You cannot, absolutely cannot think of time except by spatializing it." The temporal trajectory of our lives is given a more solid quality by mentally placing events—dates, days, weeks, months, years—next to each other. If you were to think of the last hundred years, "you may have a tendency to excerpt the matter in such a way that the succession of years is spread out, probably from left to right" (a predisposition perhaps rooted in the relationship between our left and right hand).[2] What has happened in time is "seen" (i.e., introspected) "side-by-sideness." Significantly, all languages encode the spatialization of that very fleeting, undefinable flow called temporality in the same way by relying on spatial adjectives and prepositions.

Besides time, our notions of personhood are also made spatial. As we shall see, this interiorization of the person means many things, but a common example concerns how we locate our most essential being—soul, spirit, consciousness, self, mind—"inside" the person. For moderns, this location is usually in the head or brain, while for premoderns it was often in the chest or stomach.

AIMS OF THIS BOOK: THE SHORT VERSION

In this work I demonstrate how unexpected linkages emerge once it is appreciated that historical consciousness, the sense of location, vision, and psychological processes are woven from the same intellectual fabric. Such an endeavor illuminates how our understanding of our place in the cosmos and human nature itself has been transformed through history. Allow me to state, as concisely as possible and in a list of premises, propositions, and points to be made, the aims of this book (a more detailed, nuanced explanation is found in the next three chapters).

First, human mentality is not a "black box" but a collection of constantly adapting capabilities configured by changing social conditions. Second, a crucial instance of this adaptation is the ever-increasing abstraction of time, space, and psyche. Third, this progressive "stepping back" (or meta-framing) relies on the same spatial and visual metaphors that we use to describe new notions of temporality, location, and the psyche—i.e., our ever-increasing capability to "step back" from the environment and search out its familiar features to explain the unfamiliar and generate "as if" forms of knowledge and metaphors of location and vision. Analogizing and abstracting have altered spatio-visual perceptions, expanding our introspective skills and allowing us to adapt to changing social circumstances. By showing how new introspectable faculties transformed experiences, unexpected linkages among understandings of historical change, geography, and psyche reveal themselves.

An example of this abstraction is how history stands witness to a continual "psychological interiorization." This is the fourth point. In other words, over the centuries more and more weight has been given to the "inner stuff" of the person relative to the external world. Fifth, interiorization has provided new spatio-visual experiences. An intensifying introspectable ability, interiorization helped launch the journey from a supernatural, mystical "cosmic" worldview to a scientific, measurable "scopic" reality beginning in the 1500s. Technological advances, such as the telescope and microscope, propelled meta-framing, ushering in the very modern ability to envision the astronomical vastness of the universe and the hidden world of invisible organisms. These advances demonstrate how perceptions of physical space and the internal scenery witnessed by the mind's eye are deeply interrelated.

The sixth point concerns how interiorization—sharply segregating the external from the inner world—explains why dualism has been so central to philosophical debates. Interiorization also accounts for the nineteenth-century birth of experimental Psychology that sought to measure the soul's interior. Seventh, the expansion of hypothetical, imaginary "spaces" allowed us to reimagine the passage of time as future-oriented, giving birth to our faith in political economic "progress" and the desire to relentlessly re-engineer society. Finally, history's inward turn explains why the individualistic self has come to define modernity as well as the "therapeutization" of daily life.

The Crucial Role of Psyche in History

The roaring arrival of momentous revolutions—agricultural, industrial, post-industrial—have remolded our political economic institutions and social relations like so much soft clay. Our prehistoric settlements have morphed into sprawling megametropolises, our spears have evolved into earth-evaporating megaton hydrogen bombs, our cave drawings have developed into supercomputers that process hundreds of teraflops, and our slingshots have become spaceships to the moon. And yet, we believe, throughout all the vicissitudes, variations, and vagaries of history's mighty march, an essential human identity has remained unchanged. The distinctive aspect of the human condition, often conceived as something inner and psychological, is supposedly impervious to external shocks and convulsions. This is not my view. Indeed, a primary claim of this book is that, as an adaptive force, the cultural changes of the psyche have had as much consequence as neuroanatomical modifications. This proposition, for those of us accustomed to the assumption that human nature is relatively immutable, is no doubt startling. After all, we are used to seeing our mental apparatus as fixed, unalterable, and permanent.

I contend, then, that social reconfigurations of the psyche have had as much consequence as biological evolution—i.e., that mental modifications carry as much import as bodily adaptations. Major mental shifts are not driven by indifferent physical adaptations; rather, political economic changes and sociotechnological developments propel such transformations. Consequently, cultural changes have had as much impact as genetic alterations in the human saga.

This Book's Approach: A Socially Pragmatic View of Mind

Straddling psychology and history, this project is one of those hard-to-classify interdisciplinary projects. I am interested in "big-picture" history, a type of inquiry that goes against the postmodern Zeitgeist that is highly suspicious of "metanarratives." In many such misgivings about "big history" are actually well-founded cases. Nevertheless, certain questions can only be answered when a wide-ranging and large-scale approach is adopted.[3] Changes in human mentality over the centuries are one of them. My concern is with *mentalités*,[4] a mental framework, a shared way of viewing and interacting with the world and responding to events. A group's *mentalité* is distinctive from other groups but relatively similar among members of a specific period and people. More concretely, a *mentalité* includes a repertoire of cognitive frames for understanding one's natural and social environment as well as the values and norms that guide one's behavior. The approach of Norbert Elias, who took a long-term sociohistorical perspective in his writings,[5] resonates with my interests. Two other works that are not dissimilar to my approach are Alfred A. Crosby's *The Measure of Reality: Quantification and Western Society, 1250–1600* (1997) and Charles Radding's *A World Made by Men: Cognition and Society, 400–1200* (1985).

What distinguishes my approach is how I go one step further. I offer a cultural-historical psychological analysis that challenges the view of the mind as an unchanging "black box." It is this attention to psychological processes that links this book with Julian Jaynes's *The Origin of Consciousness in the Breakdown of the Bicameral Mind* (1976). Indeed, my book continues and elaborates his thinking, and though many have explored the history of Psychology, no one has yet used his ideas to account for the very birth of Psychology and its intellectual trajectory. My goal, then, is to illustrate how concepts about the world are impacted by a socially evolving psyche and vice versa.

In order to account for how the psyche adapts to the needs of society over the centuries and the codevelopment of key concepts (time, space, the mind itself), I employ a socially pragmatic approach to mental operations. The psyche is more than mere neurophysiological processes. What are conventionally referred to as mental events should also be characterized as functions of daily interaction, intentional behavior, and sociopolitical relations. The discipline that best captures my theoretical agenda is cultural Psychology (appendix 1), or the way cultural "traditions and social practices regulate, express, transform, and permutate the human psyche, resulting less in psychic unity for humankind than in divergences in mentality."[6]

This is not a conventional history that traces, step-by-step, events in chronological order. Due to the nature of this work's agenda, I will jump back and forth between the classical, postclassical, early modern, and modern periods, ending my journey in the early twenty-first century. Originally I had intended on including a treatment of how alterations in *mentalité* relate to nineteenth-century definitions of aesthetics and fantasy as a leitmotif in pop art, as well as simulation and digitization technologies in the twentieth and twenty-first centuries. But these investigations will have to wait for another project.

Some specialists will no doubt be disappointed in superficial treatments of their pet interests and wonder why I did not include this or that topic. However, certain intellectual historical developments are sometimes not obvious unless we adopt a sweeping perspective. This work is not a comprehensive treatment, of course, and I have selected examples from the historical record that I believe illustrate salient patterns. It is a trade-off; while losing out at the level of specific details, we gain, hopefully, a more expansive vantage point that is informative in its own right. Taken separately as individual claims, some of my ideas about spatio-temporal and psychological experiences are not necessarily new to specialists. However, I feel that my original contribution lies in how I have taken subjects usually regarded as unconnected and shed light on their intertwined nature. If examples that I have neglected occur to readers that illustrate my arguments, then I will know I have been successful in convincing them of my goals.

NOTES

1. Augustine in Warner, 1963, p. 220.
2. Jaynes, 1976, p. 60.
3. But how wide-ranging and large-scale is debatable. For example, consider the work of the philosopher Ken Wilber. A reviewer strongly suggested that I incorporate some of his ideas into my book, believing that my agenda overlaps with his. But Wilber's work is grandly philosophical, far-reaching in topic, all-encompassing in spirit, and some would contend, distinctively New Age in tone. I take a less ambitious and less sweeping and more empirically grounded approach. Specifically, Wilber's use of "consciousness" is much broader than mine (e.g., see his 2007 *Up from Eden: A Transpersonal View of Human Evolution*).
4. A term used by researchers usually associated with the Annales School that focuses on the history of worldviews and daily practices.
5. E.g., *The Civilizing Process, Vol. I, The History of Manners,* 1969, and *The Civilizing Process, Vol. II, State Formation and Civilization,* 1982.
6. Shweder, 1990, p. 1.

Chapter Two

Purposes and Premises

Tracing the Trajectories of Human Experience

When I was about seventeen, I discovered a new space. While passing by a prep room in my family's funeral home business, I noticed a corpse that had been autopsied. Till this day it was not the bloody inner cavity, the missing organs, or the deathly pale lifelessness of what lay on the table that left a lasting impression. I had seen and moved deceased individuals before, some grossly disfigured, while helping out at the business. What struck me was a new spatio-visual experience: The realization that humans possess not only width and height, but also depth. In other words, for the first time in my life I saw a person, his rib cage emptied of its viscera and hollowed out like a pumpkin, as a three-dimensional being. This experience left me wondering if people have known spatiality differently depending on historical period. But it also made me consider the relation between perception and socialization, what happens to our definitions of space when our environment reveals new places, and the linkages between physical space and what occurs in our minds.

KEY PREMISES OF THIS BOOK

Macrocosm, Microcosm, and Introcosm

No matter what period we are discussing, we can speak of a common and time-honored cosmic split into two different types of space: macrocosm and microcosm. The former means the world on the grandest scale, while the latter refers to an individual (or a community) conceived as a miniature world or universe. The terms, macrocosm and microcosm, are in fact quite imprecise and not terribly useful unless a particular issue is being investigated. Nevertheless, their very vagueness provides us with a useful conceptual starting point. Allow me to narrow their definitions somewhat. Macrocosm refers to either the world experienced close at hand (terrestrial) or envisioned as a distant reality (celestial), in all its grandeur, strangeness, and terror. It encompasses spaces and places immediate and mundane, as well as remote and extraordinary. Microcosm, on the other hand, is more person-centered, and though it certainly concerns spatial elements, it may designate not just the person but also what is "within" the person.[1] According to ancient cosmologies the macrocosm and microcosm reflected and influenced each other. "The old notion of the microcosm and the macrocosm, the physical world as a double-reflecting mirror between our psyches and a transcendental other world, is more deeply embedded in us than we imagine."[2] Not surprisingly, changes in spatial conceptions about the macrocosm led to parallel changes in the microcosm and vice versa (chapter 7). The linkages, being subtle, are not easily discerned, and occurred over a time span of centuries.[3]

Three Worldviews

To appreciate the aforementioned interlinkages, I frame my arguments within three major worldviews or *mentalités* of spatio-visuality. Since the earliest days of civilization, we have had to learn these *mentalités* in order to adapt to the vagaries of historical change. They concern how we "visualize" (whether with the real eye or the mind's eye) and experience placeness. To appreciate visualization (again, whether of the physical world or mentally), spatiality must be appreciated as it delineates the boundaries of what is seen.

The first *mentalité* was composed of the microcosm (one's body and immediate surroundings) and macrocosm (the expansive physical world). It has its roots in prehistory and lasted until about 1000 BCE. During this epoch time was viewed as cyclical, and the past was an idealized "Golden Age." Space was finite, bounded, concrete, and heterogeneous in character. Nature was thought of in terms of inherent qualities; the notion that the world's stuff came in quantities certainly existed, but measurement was primitive by modern standards.

The second *mentalité*, which emerged after 1000 BCE, was more or less a continuation of the first, except for one enormous difference: the introduction of a new space—indeed, a whole new inner universe—within the person. This spatialization of psyche, or metaphorically positioning an introcosm within the individual where decision-making transpires, is a primary feature of conscious interiority (chapters 3, 5, and 6). This cavitation can be described as a hollowed out mental place modeled after the perceptuo-physical world. The social construction of a psychoscape allowed people to experience a quasi-sensory, introceptive world in which dwells a self.[4] The introcosm, then, was a counterpart to the microcosm and macrocosm. The reasons for this expansion of mind-space and privileging of the internal life of the individual are not obvious unless a long-term, marcrohistorical perspective is adopted that appreciates general trajectories and wide-ranging global patterns.

The third *mentalité*, which began to take shape from approximately the 1500s, witnessed the introduction of technologies that radically altered the way we see the world. Specifically, the telescope and microscope, like sunshine illuminating dark rooms and the dank dungeons of superstition, began to dispel the numinous "cosm" from our experiences and mappings of reality. These instruments took both the invisibly miniature and the unimaginably grand in scale and transmuted them into something seeable and measurable, i.e., "scopic." Other scientific developments began to alter our view of nature; the cosmos began to transition from a mysterious gathering of unpredictable powers to a collection of objects amenable to categorization, measurement, quantification, and mathematicalization.[5] People started to visualize the "stuff of reality as aggregates of uniform units, as quanta: leagues, miles, degrees of angle, letters, hours, minutes, musical notes."[6] The macroscopic and microscopic perspective, due to impressive techno-scientific advances in the nineteenth century, would generate even more spatialization, thereby momentously modifying our worldview. A technologized literacy (i.e., spread by Gutenberg's printing press from the mid-1400s) allowed readers to virtually visit and imagine new places and no doubt expanded the individual's abstracting, hypothetical, and speculative abilities (at least among the reading population). Technology, its social usage, and cultural interpretation are very difficult to disentangle. For instance, John Crary notes how for Newton and Locke the camera obscura[7] was a "model simultaneously for the observation of empirical phenomena *and* for reflexive introspection and self-observation."[8] In any case the point is that advances in technoscience accompany changes in the mind itself.

Meanwhile, at the psychological level, the supernaturally defined introcosm became the introscope, as more and more attention was devoted to introspection in the nineteenth century in the emerging discipline of Psychology (chapters 11 and 19). Indeed, introspection would become a means to an end for dissecting the inner person, as Psychology shifted from philosophical self-reflection to an instrument-based, objective observation of the psyche.[9] Table 2.1 elaborates on some of the aforementioned themes that constitute the arguments of this book.

Metaphors of Mind

So what is the mind? How should it be studied? Below I provide the context for the following premise that undergirds this book's arguments. Let us begin

Table 2.1. Traits of Cosmic and Scopic Worldviews

	Cosmic Worldview	Scopic Worldview
Spatio-Visuality	Macrocosm, Microcosm, Introcosm	Macroscopic, Microscopic, Introscopic
Time	Cyclical: Past–Present	Linear: Past–Present–Future
Space	Concrete, Heterogeneous, Bounded, Finite	Abstract, Homogeneous, Unbounded, Infinite
Person	Soul: Supernatural Entity	Mind: Secular–Scientific Entity
Interpretive Mode	Nature Constituted by Qualities, Powers, Imprecise Essences, Traits	Nature Understood as Quantities, Forces, Measurable, Mathematicalized
Subject–Object	Subject & Object Vaguely Distinguished (e.g., soul and body not clearly differentiated)	Subject & Object Clearly Distinguished (e.g., mind–body dualism)

with two metaphors of psyche: mind-as-vase and mind-as-putty. In the former trope, society deposits its cultural contents into a predetermined container, and it is the configuration of the container that shapes the cultural stuff. To conventional Psychologists, socialization and enculturation are of secondary concern, since it is the shape of the vase—i.e., the "central processing mechanism" (CPM)—that matters. Thus, Psychology becomes a search for psychic unity and psychic structures, since it is assumed that though cultural information may vary historically and from place to place, the basic shape of the container is the same for everyone. The mind is a receptacle into which society pours its values, ideas, concepts, classificatory schemes, fears, aspirations, etc.

For the mind-as-putty metaphor, the focus is on the socially constructed aspect of mind, which can be molded into a number of shapes for culturally determined purposes. The "putty material" is the same, of course, for all individuals, but it can be formed into radically different configurations depending on the circumstances.

Of the two, the vase trope is the more commonly accepted theory and has widespread purchase among many research Psychologists. The putty trope, in my opinion, is more useful and captures better the workings of mind *if one is interested in the psyche as shaped—but not determined—by historical forces*. It is intended to illustrate that though we are all born with the same neurological apparatus that provides very general, undefined, and diffuse mental capabilities, these facilities amount to very little unless the individual is socialized. Rather than only searching for universals, I argue, Psychology should also be searching for psychic plasticity and psychic diversity.

Though the container metaphor is in some respects a gross exaggeration, it is useful for illustrating how some thinkers afford too much significance to the shape of the container and not enough to the cultural stuff with which it is filled. If the container metaphor were correct, then socialization would not be as necessary as it apparently is for constituting human nature. Also, if the container metaphor were correct, more uniformity through history and across different cultures would characterize the human psyche. The various shapes of putty that appear to be the same across different societies are due to similar sociopolitical forces impacting on the putty, not necessarily universal structures inherent in the putty.

This contrasting of the vase and putty metaphors, of course, is in a sense just an expression of the form/content debate.[10] In any case, not a few researchers assume that form is more fundamental than the messy cultural stuff added on, placed in, and recorded by the mind. However, the content is as important as the form for coming to terms with psychological processes. Or perhaps we could say that there is no easy way to tease form and content apart. Enculturation, socialization, outside influences, and other stuff of the real world—that for many seem to dirty the crystalline and neatly formal view of mind advocated by some cognitive scientists—cannot be readily extracted from mind.

Implications and Significance

Why is it more profitable to regard the psyche as highly plastic? I give two reasons. First, a careful investigation of psychohistory and comparative psychologies reveals that the psyche does vary according to time and place. However, because we are naturally drawn to similarities, we are easily distracted from the many significant differences between ethno-mentalities. Thus, we often fall into the trap of assuming that the psyche is and has been a mere container filled in by the vicissitudes of period and place. Consequently, it is easy for many historians to view the mind as a black box.

A second advantage of the mind-as-putty view can be appreciated by introducing another pair of contrasting metaphors: mirror or tool. Some researchers have privileged the former—i.e., the psyche reflects not just the world (represents the external physical environment); it also operates in such a way that we can mentally "see" reasoning (represents internal rational processes). In the mind-as-tool trope, however, the psyche adapts to and functions in society. It processes information in order to accomplish tasks rather than passively reflecting the environment. The "structures of mind are not so much the imperatives of culture as its implements."[11] The psyche, I contend, is not mirror-like; rather it is a malleable, multipurpose tool configured by changing social conditions. Though its plasticity has limits, it can be refitted for purposes of adaptation (social pragmatics). The mind cannot be bracketed off from society. Nor is it a mysterious substance dwelling in our heads. Once this is understood, we can look for mind where it occurs: in the churning of sociohistorical change.

AIMS OF THIS BOOK: THE LONG VERSION

The Intertwined Fabrics of Space, Time, and Psyche

I make my arguments by exploring the three realms with which we are all intimately acquainted: place, the passage of moments, and the interior person. Here I will state my thesis as briefly as possible: I investigate how these spheres of experience at bottom are intimately interrelated. We conceptualize the drift of events, the crooks and crannies of the cosmos, and personhood using the same primary, unifying notions: spatiality and visuality. Changing definitions of time, space, and the essence of personhood follow parallel historical trajectories since they all mutually impact each other. Metaphors of space and vision constitute the cloth from which these three realms are woven. Follow the threads of one, and they lead you to the stitchings of the other two. The *particulars* of how these three spheres of human experience have influenced each other are not always obvious, but the *basic metaphoric process* of how they are interconnected is actually not very difficult to appreciate.

Mental Space, the Linearization of Time, and the Invention of the Future

Societies, realizing how powerful the metaphor of placeness was, applied it to the progression of days, the sequence of seasons, and the journey of life. But such temporal successions were for the most part cyclical and closed. And though in fifth-century BCE Greece Herodotus began to write something like history, it would be centuries until time was linearized and regarded as open-ended. It would not be until the eighteenth century that history would be conceived not as endlessly spinning circles, but rather as an arrow, following a straight line to an uncertain but hopeful time to come.

Conceiving time as progressing in a straight line is made possible by the spatialization of psyche, by hollowing out a pretend cavity in our heads where we "see" time (or at least somehow feel it in some indescribable way) as an unfolding sequence of events in linear fashion. And this mentally lining up past or unrealized events side by side would eventually lead to one of the most significant social innovations in human history: the future. The import of the yet-to-be for human societies lies in how it greatly enhances our ability to strategize, assess risk, devise escape plans, better apportion unexpected surpluses, and in general organize more efficiently and adapt to changing conditions. At the economic level, belief in progress means "becoming rich," and acquiring the comforts of the middle class (embourgeoisement). The future, this imaginary locale, which is related to our almost religious faith in "progress,"[12] was invented roughly by the early nineteenth century.

From an individual perspective, progress corresponds with "self-improvement," "self-development," and "personal growth," ideas that resonate in our therapeutic society (chapters 20 and 21). Our pursuit of the future has impacted politics, economics, sociology, psychology, science, and technology. Desiring after this imaginary place has opened up potentialities and possibilities, while whetting our appetites and expanding our expectations for a better world (Part III, Time: Modern Millenarianism and Politics as "Progress" and chapter 17).

The Spatialization of Space

Claiming that we employ space to conceive space smacks of circular reasoning. However, as noted above, new technologies, in particular the telescope and microscope, introduced astronomical and miniature scales of vision that were simply unimaginable before the 1600s. The meaning of "very far" and "very small" would never be the same. The astroscopic would stretch the cosmos like an ever-expanding giant balloon. And at the same time, the microscopic (and eventually the nanoscopic) would collapse the cosmos, squeezing it like a sponge, shrinking it down like unfortunate Scott Carey in the 1957 sci-fi movie *The Incredible Shrinking Man* or the miniaturized crew of the *Proteus* in the 1966 *Fantastic Voyage*.

As science progressed, it would need to spatialize into existence more spaces in order to capture the difficult-to-represent and dizzying abstruseness of advanced mathematics, non-Euclidean geometries, and other superabstract notions. But it was the body itself that, by the 1500s, was opened up, described, and depicted, thereby revealing a new medical spatiality. We can thank pioneers such as the Flemish physician Andreas Vesalius (1515–1564) for helping found modern human anatomy. In more recent times, digital technologies and supercomputers have afforded us the magic of simulation as well as entire cyberspatial universes. These developments will in the distant future undoubtedly rival science fiction in how they modify our neurocultural apparatus.

Science, after all, is about searching for new places, of tunneling for new truths, mining the guarded treasures of Mother Nature for secrets, and carving out novel realms of experience (Epilogue, Visualizing New Vistas of Modernity and Self-Hood). Electron and scanning probe microscopes open up hidden worlds invisible to the naked eye. Radio and spaced-based telescopes pierce the vast depths of star-speckled space. Functional magnetic resonance imaging (fMRI), magnetoencephalography (MEG), electroencephalography (EEG), positron emission tomography (PET), and near-infrared spectroscopy (NIRS) attempt to map the brain (as if the self is really "in" the head somewhere). What is important is that no matter which era and its concomitant technology we examine (from the earliest writing, printing, computerized, digitized, to the information-suffused), technology and spatialization mutually reinforce each other, creating systems of accelerating complexity that produces new vistas that would boggle the minds of our predecessors.

Conscious Interiority and the Social Construction of the Self

Though we may assume it is natural to metaphorically locate thoughts "inside" an individual, this is a learned cultural convention, not an inborn biological capability. This interiorization of the person is historically much more recent than is commonly believed. Surely the fact that before the first millennium BCE individuals rarely recorded their own subjective experiences should force us to wonder if they possessed conscious interiority. As full-blown psychological interiorization gathered steam about three thousand years ago, it brought with it a train of new cognitive abilities that together constitute conscious interiority (chapter 4) (see appendix B for a treatment of what it is *not*).

Intellectual revolutions of great depth and magnitude have deep historical roots. To understand these tectonic shifts, whose mountain-moving impact was in full effect by the 1800s, we need more than just a list of the usual suspects of "isms"—industrialism, capitalism, rationalism. Despite the impact of these salient ideologies, the time frame of these master narratives is historically limited. Therefore, a long-term concept that illuminates even larger sociopolitical patterns is required: interiorization.

In more modern times certain features of conscious interiority have been augmented. For example, individuation, or paying keen attention to what makes us each a unique person, is evident in practices focused on "finding one's self" since the 1960s (though the roots go back to early-nineteenth-century Romanticism) (chapter 16). Our current obsession with self-disclosure (aided by Facebook and other social media technologies) would have been frowned upon not that long ago. But now we think nothing of sharing our random thoughts with an anonymous audience. This is an example of interiorization—i.e., privileging the private contents of consciousness. And a key feature of conscious interiority, self-reflexivity, undergirds a salient trend in modern politics: increasing interest in "self-identity" and "identity politics" ("collectivizing the individual," i.e., rather than joining a movement, equating one's self with a racial, ethnic, or other collectivity in order to make oneself feel more important) (part IV, Self: Turning the World Inside Out).

NOTES

1. Admittedly, this conceptual formatting is more or less inspired by the Western tradition. However, I contend that, due to the very nature of the human condition and the nature of space, it strongly resonates with other civilizations and cultures.

2. Nelson, 2001, p. 6.

3. Less appreciated is how subjective sceneries witnessed by the mind's eye are projected back onto the outside, physical world. It is not hard to see how we can move from "real physical" to "pretend metaphorical" space, but it is difficult to see how metaphors configure our view of the literal. One example: Victoria Nelson's use of "psychotopography": "A psychotopographer is the artist who devotes herself to describing—with varying degrees of awareness about the true nature of the subject—the images of these inner regions as she discovers them in an imagined exterior landscape. Working backward from the sum of these details, the reader gains a picture not of what lies without but of what lies within" (Nelson, 2001, pp. 110–11).

4. I call such processes "introception" (in contrast to perception). A more prosaic expression might be introspection, but strictly speaking, this latter term ("seeing within") is restricted to the "mind's eye," while introception encompasses all quasi-perception, such as hearing music "in our heads."

5. See Alfred W. Crosby, 1997, *The Measure of Reality: Quantification and Western Society, 1250–1600.*

6. Crosby, 1997, p. 10.

7. An optical device that projects an image of its surroundings on a screen. Used in drawing and for entertainment, it played a role in the development of the camera. A hole would be made in a box or room so that light would pass through it and strike a surface inside where it is reproduced, inverted, with color and perspective preserved.

8. Crary, 1994, p. 40 (emphasis in original).

9. Fuchs, 2000.

10. Or what philosophers of mind refer to as the rationalist-versus-empiricist debate.

11. Sahlins, 1976, p. 122.

12. See Bury's important *The Idea of Progress,* 1932, and Nisbet's *History of the Idea of Progress,* 1980.

Chapter Three

The Magic of Metaphors

How Our Minds Make the World

In his sensitive exploration of the evolution of English words and how their changing semantics reflect historical conditions, Owen Barfield noted that the birth of a historical sense and our modern self-consciousness may seem to have little connection with one another. However, a "larger process" links these two developments. Barfield calls this "internalization," or the shifting of the center of "gravity of consciousness from the cosmos around us into our beings."[1] I share Barfield's intuition about this grand interconnectedness, and in this work I explore how mental activities, new notions of time, and the discovery that space could be greatly expanded or shrunk to an unimaginable degree are intimately interrelated. The theoretical key to unlocking their relations is the ability to mentally "step back" and look out on (whether "in" one's head or in physical space) new frames of conception. This endless reframing cannot operate without key metaphors of space and vision. In this chapter I discuss the theoretical specifics of how this meta-framing occurs.

META-FRAMING TIME, SPACE, AND PERSON

When confronted with a challenging situation, a basic human mental ability is to stand back, assess the context, extract crucial specifics, isolate out key details, and conceptualize an appropriate response. If conditions change in reaction to our response, we again step back and adopt a bird's-eye view, abridging, condensing, and summarizing newly formed ideas. There is no need to take in all the aspects of the situation; indeed, details can get in the way. Picking out what is relevant is crucial. This continuous reframing of our circumstances can be referred to as abstraction, which not only is the basis of our daily navigations through life but also characterizes the hypothesizing and theorizing behind our grandest intellectual achievements as a species.

Reframing (or re-picturing) our circumstances can be expressed in any number of terms, many prefixed with "meta," meaning "beyond," "about," "after," "beside," "with," "between," or "among." This prefix is employed in a host of terms having to do with taking an analysis to another level of abstraction, as in metaphysics, meta-cognition, meta-representation, meta-referentiality, or meta-discourses ("second order discourses that attempt to reflect on first order discourses").[2] These terms capture the aforementioned procedures of metaphorically "stepping back" to get a better view. In Psychology meta-memory indicates people's knowledge about whether or not they would remember something if they attempted to remember it. Meta-emotion means people's emotions about their initial emotions (or somebody else's basic emotion). Linguists refer to grammar as a meta-language—i.e., a language for describing a language. Meta-data are data about data—i.e., the circumstances behind the collecting of data, who produced them, why, when, etc.

Consider another related concept: Meta-pictures. Mitchell defines these as "pictures about pictures—that is, pictures that refer to themselves or to other pictures, pictures that are used to show what a picture is." Mitchell suggests that the function of a meta-picture is to "explain what pictures are," to "stage as it were, the 'self-knowledge'" of pictures. Such "self analysis" is "directed not only at the medium, but at the determining conditions of the work—its institutional setting, its historical positionality, its address to its beholders."[3]

Another term that summarizes some of these aforementioned processes is "metaization." This ungainly but useful term is borrowed from Wolf Werner, who defines it as the "movement from a first cognitive, referential, or communicative level to a higher one on which first-level phenomena self-reflexively become objects of reflection, reference and communication in their own right."[4]

The meta-prefixed words mentioned above actually possess different nuances, but for my purposes I require a term that includes the various senses of the previous examples but is more general. My own neologism accommodates elements of the previous definitions: meta-framing.

When viewed over the *longue durée* and as a general process, meta-framing can be understood as an advancing, accumulative, and progressive development. For centuries societies have been amassing facts, collecting evidence, enumerating principles, and cataloguing theories. But this ever-growing storehouse of knowledge is not a mere list of bits of disconnected information. Indeed, more than just the sum of its parts, the overriding characteristics of meta-framing are overarching generalization, sweeping supposition, and grand if dry intellection. In a word, meta-framing relentlessly pursues an ambitious abstractness as it scales a transcendent, towering vantage point from which everything can be objectively surveyed. This never-ending climbing up to higher and higher vantage points—ceaselessly ascending to God's viewing platform—allows the conjecture, speculating, and reasoning that propels techno-scientific advances forward.

FROM COSMIC TO SCOPIC SPACE

As an example of meta-framing as a long-term historical process, consider our evolving definitions of space. As the cosmic worldview collapsed due to the discoveries introduced by the telescope, microscope, and the genius of Kepler, Copernicus, Galileo, and Newton, a scientific, technologically enhanced "scopic" worldview pushed aside anthropocentric, human-scaled space. Scopic means utilizing instruments to measure, count, and quantify Mother Nature (and by the late nineteenth century, society and the psyche would, respectively, come under the magnifying glasses of sociology and Psychology). The meta-framing of the stuff of the cosmos represents a change from a classical-medieval to a modern mentality. In a sense, the discovery of places that cannot be directly perceived (i.e., the telescopic, astronomical, microscopic, and more recently, nanoscopic) constitutes more of a qualitative (a different stuff) than a quantitative change (more of the same stuff) in our understanding. To summarize all this: A *cosmic worldview* (macrocosmic, magico–astrological, microcosmic) evolved into a *scopic worldview* (macroscopic, tele-astronomical, microscopic).

Depending on the historical period and the confluence of religious thought, science, and technologies, different types of spaces come into existence or disappear, delineating the vagaries of person and cosmos. Modern technologies, via meta-framing, have constructed new imaginary spaces, such as cyberspace, which is metaphorically external but visible to the person. And modern Psychology has given us a depth that is internal and invisible to the person, such as the unconscious. For example, during the European Middle Ages Heaven and Hell were considered real spaces, external to the person and, depending on the context, sometimes invisible, sometimes visible. By the twentieth and twenty-first centuries, the relocation of vision to a plane "severed from a human observer"—computer-aided design, synthetic holography, flight simulators, computer animation, robotic image recognition, ray tracing, texture mapping, motion control, virtual environment helmets, magnetic resonance imaging, multispectral sensors—indicated a "relentless abstraction of the visual." This is an instance of meta-framing, so that "visual images no longer have any reference to the position of an observer in a 'real' optically perceived world."[5]

THE WORLD-SHAPING POTENCY OF METAPHORS

The prefix of "meta" means "something (X) goes beyond/comments on/reflects upon itself (X)." However, an extended, complex function of meta-framing is metaphor, i.e., something (X) describes something different (Y). Much is written about this conceptual subspecies of meta-framing and numerous definitions exist: Metaphors transform the strange into the familiar[6]; they allow us to see "something from the viewpoint of something else"[7]; they are figures of speech whereby "we speak about one thing in terms which are seen to be suggestive of another."[8] Though conventionally understood as a rhetorical device used in poetry and colorful prose, metaphor is actually the foundation of literal language and key concepts.[9] Language "climbed up from the concrete to the abstract on the steps of metaphors," and abstract words are "ancient coins whose concrete images in the busy give-and-take of talk have worn away with use."[10] Thinking, then, should be considered a collection of metaphors shaped by history.

How Metaphors Shape Worldviews[11]

In order to appreciate the magic of metaphor, let us propose two types of tropes: (1) *figurative metaphors* that describe how an individual interprets an analogical experience as a figure of speech; and (2) *literal metaphors* that describe how an individual interprets an analogical experience as being literally true.[12] For example, we do not believe that internal organs, such as the heart, constitute the mind. But for the ancient Greeks, such figurative descriptions were taken literally—i.e., viscera contained or in some sense *were* what we would label psychological entities. In other words, via the process of literalization the heart, when utilized in discussions of mentation, was *not* a mere figure of speech.[13] Literalization shapes worldviews or definitions of key concepts, such as time, space, matter, and person, in subtle and unforeseen ways, forming the basis of "master," "foundational," or "root metaphors" that become incorporated into a society's undergirding assumptions that constitute its worldview.[14]

A metaphor's persuasive power comes from its two components working together: the unknown thing to be described (metaphrand) and the known thing doing the describing (metaphier). For example, in the expression "time flies like an arrow" time is the more abstract, difficult-to-conceive metaphrand, while arrow is the more concrete, easier-to-grasp metaphier. Note, however, that it is not really the arrow itself that provides the descriptive impact, but something about the arrow—its unstoppable speed, linear preciseness, determined directionality—that imparts explanatory power. These secondary, ancillary meanings can be called paraphiers. A metaphier, then, possesses associated paraphiers which become the paraphrands of the metaphrands. So it is the network of meanings in which a concept is embedded that grants expressive power to figures of speech (though not all metaphors are particularly generative).[15]

As an example, consider the tropes all languages employ to describe mental events. In order to make the intangible, unobservable, and inner happenings of our minds take form, the spatio-visualization of mental phenomena is an indispensable metaphoric technique. In English the "most prominent group of words used to describe mental events are visual. We 'see' solutions to problems, the best of which may be 'brilliant,' and the person 'brighter' and 'clear-headed' as opposed to 'dull,' 'fuzzy-minded,' or 'obscure' solutions." In our mind-space "we can 'approach' a problem, perhaps from some 'viewpoint', and 'grapple' with its difficulties, or seize together or 'comprehend' parts of a problem, and so on." And consider the adjectives we rely on to describe physical behavior. These are borrowed from real space: We

> speak of our minds as being "quick," "slow," "agitated" (as when we cogitate or co-agitate), "nimble-witted," "strong-" or "weak-minded." The mind-space in which these metaphorical activities go on has its own group of adjectives; we can be "broad-minded," "deep," "open," or "narrow-minded"; we can be "occupied"; we can "get it," let something "penetrate," or "bear," "have," "keep," or "hold" it in mind.[16]

As in physical space something can be at the "back" of our mind, in its "inner recesses," or "beyond" our mind, or "out" of our mind. In arguments we try to "get things through" to someone, to "reach" their "understanding" or find a "common ground," or "point out," etc.[17] Figure 3.1 portrays how the spatio-visualization of the psyche eventually led to dualism (this is discussed in more detail later).

HOW MIND-WORDS ARE REIFIED VIA METAPHORS

Why are metaphors so persuasive? Victor Turner's attention to how the sensory and ideological associations of a symbol mutually reinforce each other has profound implications for understanding how language constructs mind. His insightful work into the dynamics of symbolism aids us in seeing why metaphoric imagery is so effective. For Tuner a symbol (such as a flag or religious icon) is associated with certain perceptions that, by their physical qualities, evoke physiological and emotional reactions within the individual. But this same symbol denotes social norms and ideas.[18] In other words, what is important to us draws their efficacy from the interchange of qualities between the organic and the cultural, with society (more specifically, idioms and linguistic practices) as the matrix in which this interchange occurs. In-

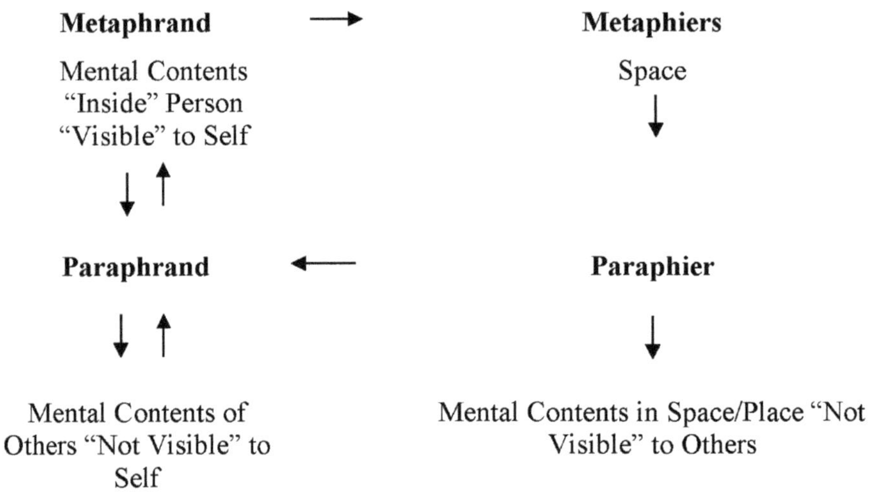

Figure 3.1. Transition from a Cosmic to Scopic Worldview.

deed, thinking in general can be analyzed as operating through the cultural, social, and organismic aspects of human experience.

The persuasive power of metaphors, like symbols, is also derived from and grounded in a process in which ideas and experience are exchanged. Culture, as information, is the abstract, ideational aspect of the process. Taken from the observable and tangible world, metaphors, through the medium of social behavior, describe and define sensate and visceral experiences. The experiential part of the process occurs within the organism, and its sensations confirm and reinforce the cultural, ideational part of the process. This is an ongoing, continuous process, with the ideational and the experiential interlocked in a mutually reinforcing dynamic with social activity as the mediating practice (figure 3.2).

The use of bodily parts and sensate experiences to describe mental activity is a linguistic universal that follows a definite historical trajectory, as elucidated by Jaynes:[19] *Physically experienced bodily parts* (metaphiers—heart, head, stomach, liver, blood) describe *psychological processes and events* (metaphrands—cognition, emotions, feelings, volition).

THE METAPHORS OF SPATIALITY AND VISUALITY

Given the great diversity of and historical changes in cosmologies, do global patterns of spatiality and visuality exist? Is it possible to discern a basic common trajectory that different cosmologies all share? Can a basic language

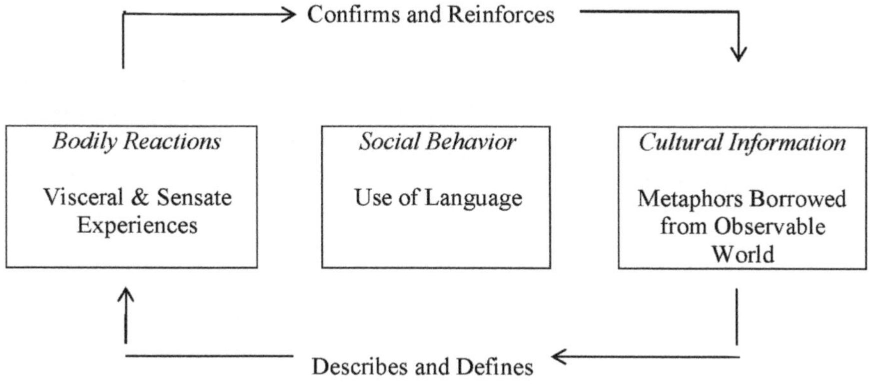

Figure 3.2. Spatio-Visual Metaphors of Mind.

describing the interplay among person, cosmos, and space be delineated? What would such a universal idiom, transcending historical periods and locale, look like?

We can begin by using two axes that plot space and visibility: "External-versus-internal" (vis-à-vis an individual's body) and "visible-versus-invisible" (from an individual's perspective). Spatiality and visibility, which are useful starting points since they are universally recognized, may be understood either literally or metaphorically; or phrased differently, these axes plot different types of space: real or imaginary. For example, what is visible and external to a person occupies real space. But certain types of real spaces do exist that are invisible to the naked eye, i.e., at the microscopic or on the astronomical scale. Then there are imaginary places, such as mental space, which are metaphorically "visible" (via introspection) and "internal" to the person (i.e., one's psychological processes do not really occur "in" one's head) (appendixes C and D).

THE PRACTICAL POWER OF METAPHORS

Metaphoric maneuvers are not a just trick of language or a poetic device with merely aesthetic value. They do more than depict; they alter our very definitions of reality. Their significance lies in how we use them to conceptually cope with and adapt to ever-changing social and natural reality. Metaphors are not descriptions; they govern our perceptions and shape our practical responses to the world around us. Metaphor is not "only a form of speech, but more fundamentally a form of thought, having basic epistemological functions."[20] It is not simply descriptive, but transformative.[21] Indeed, "explicit self-consciousness seems to be the result, not simply the progenitor, of metaphorical thinking."[22]

Metaphors, then, are attempts to make the incomprehensible more understandable, to solve a problem, to adapt to changing circumstances, both mundane and trivial as well as momentous and substantial. They produce "truth," which according to Nietzsche, is a "mobile army of metaphors" (though we should add the caveat that metaphors can mislead as well as illuminate). Change the dominant metaphors, and the mental paradigm changes. But dominant metaphors change in accordance with political, economic, and religious transformations. Figure 3.3 illustrates how the spatialization of time generated more elaborated meanings. But such significations are not airy speculations disconnected from the real world. For example, circular notions of time reflected the understanding of premodern societies that they could perish if agricultural productivity was not adequate. Cycles of birth–death–rebirth corresponded with natural rhythms and periodicity. Civilizations that followed the comforts of predictable and predetermined tempos, regularities, and patterns saw little need for radical innovation, whether of technology or political economic institutions. "Efforts to improve society . . . seem quite pointless if everything is assumed eventually to return to the same state."[23] But with steam- and electricity-driven industrialized production of a cornucopia of commodities, time became a path to open-ended and nonfatalistic progress, a noncyclic and nonpredictable road to a better future. Time was metaphorized into a "direction to be followed": The passage to time (metaphrand) was described by space/place (metaphier), which in turn generated the paraphiers of a trail, path, road, river, or a journey to new locations. These tropes then became the paraphrands of pursuing personal goals or fulfilling one's ambitions, moving ahead in life, encountering new experiences.

HOW METAPHORS AID META-FRAMING

Characterizations of macro- and micro-spaces rely on "spatial descriptions." This is obvious, but I mention this to point out how such descriptions are not metaphoric (which means "something [X] describes something different [Y]"). Beginning in the seventeenth century the micrographic and macrographic accounts generated by new instruments, then, should be called instances of meta-framing—i.e., "something refers back upon itself but in a novel way" (in this case, spatiality). However, metaphors of mind enabled the meta-framing of space (and for that matter, time and the psyche). In other words, changes in conscious interiority greatly facilitated the transformation in understandings of time and space since about 1600. Why? Because metaphors that expanded our introspectable psychoscapes made it much easier to

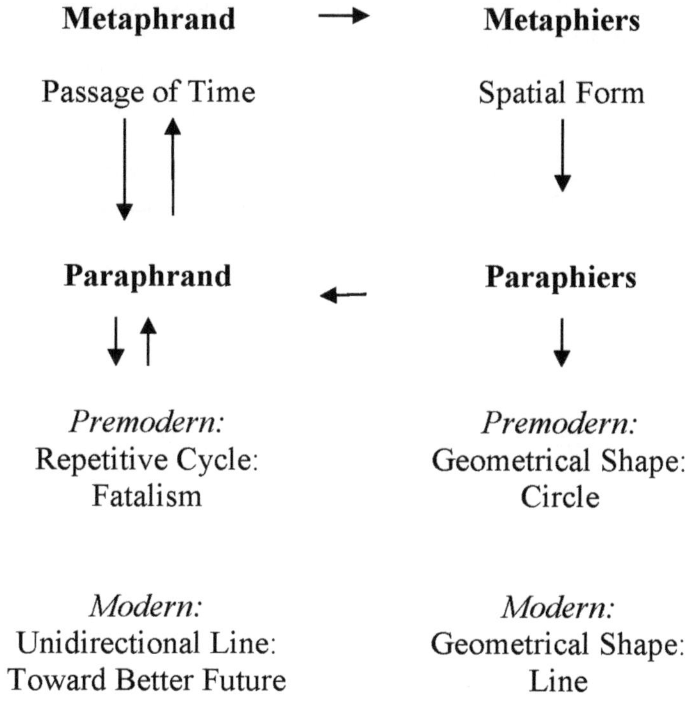

Figure 3.3. The Metaphoric Spatialization of Time as Geometrical Shapes: Premodern Societies.

imagine locations hidden from the naked human eye. Such whereabouts—vast distances and infinitesimally small places—are from a practical perspective "as if" spaces whose existence becomes more convincing when we are able to experience imaginary inner terrains. The enlargement of subjective sceneries and the mapping of new geographies of the cosmos are two sides of the same coin. And the reconfigurations of the psyche are an adaptive force in their own right.

AN ILLUSTRATION: THE COSMOLOGY OF HUMORS

As an example of how metaphors tie together macro–micro–introcosmic complexes, consider how "humors" (from the Latin *umor,* meaning "moisture" or "fluid") linked together natural phenomena, physiology, pathology, and Psychology. The metaphrands of dispositions and temperament were described by metaphiers: courageous, hopeful, amorous (sanguine) by blood, liver; easily angered, bad tempered (choleric) by yellow bile, spleen; despondent, sleepless, irritable (melancholic) by black bile, gall bladder; calm, unemotional (phlegmatic) by phlegm, brain/lungs. These metaphiers generated paraphiers (warm and moist; warm and dry; cold and dry; cold and moist) which described the corresponding paraphrands of spring/air; summer/fire; autumn/earth; winter/water. These in turn related back to the original metaphrands.

Though the theory of humors has become associated with the great physician Galen (129–c. 200), it can be traced back to the ancient Greeks. It was believed that four "cardinal humors," in addition to corresponding with cosmic forces and features, described the characteristics and dispositions of individuals. The theory of humors lives on in modern usage: We describe personalities as sanguine or melancholic and discuss each other's "temperament" (from "bodily temperature").

To summarize so far: This book details how three realms of existence—the person as a container of psychological events, new understanding of temporality, and the encountering of new places and locations—were conceptually interlinked by using metaphors and meta-framing.

NOTES

1. Barfield, 1967, p. 171.
2. Mitchell, 1995, p. 35.
3. Mitchell, 1995, p. 35, 36.
4. Werner, 2011, p. vi.
5. Crary, 1994, pp. 1, 2.
6. Ozick, 1986, p. 67.
7. Brown, 1977, p. 77.
8. Soskice, 1985, p. 4.
9. Indeed, the term "metaphor" itself is metaphoric. It comes from the Greek *metaphorá* (*μεταφορά*) meaning "transfer," but this latter word originates in the more concrete *metaphero* (*μεταφέρω*) indicating "to carry over." To go one step further, *metaphero* can be analyzed into *meta* (*μετά*) for "between" and *phero* (*φέρω*) meaning "to bear," "to carry."
10. Jaynes, 1976, p. 51.
11. As the number of works on metaphor is overwhelming, I make no pretense of offering a comprehensive literature review. But some works that I have found particularly useful include: Barclay, 1997, "The Metaphoric Foundation of Literal Language: Towards a Theory of the Reification and Meaning of Psychological Constructs"; Billow, 1977, "Metaphor: A Review of the Psychological Literature"; Haskell (ed.), 1987, *Cognition and Symbolic Structure: The Psychology of Metaphoric Transformation;* Gendlin, 1998, *Experiencing and the Creation of Meaning: A Philosophical and Psychological Approach to the Subjective*; Johnson (ed.), 1981, *Philosophical Perspectives on Metaphor*, and 1987, *The Body in the Mind: The Bodily Basis of Meaning, Imagination, and Reason;* Lakoff and Johnson, 1980, *Metaphors We Live By*; Leary, 1990a, *Metaphors in the History of Psychology*, and 1990b, "Psyche's Muse: The Role of Metaphor in the History of Psychology"; MacCormac, 1976, *Metaphor and Myth in Science and Religion*, and 1985, *A Cognitive Theory of Metaphor*; Ortony (ed.), 1979, *Metaphor and Thought*; Shibles, 1971, *Metaphor: An Annotated Bibliography and History*; Smith, 1985, "The Metaphorical Basis of Selfhood"; Van Noppen, De Knop, and Jogen, 1985, *Metaphor: A*

Bibliography of Post-1970 Publications. See also Gibbs, 1994, 1999, 2003, and Gibbs and Steen, 1999.

12. McVeigh, 1996. Figurative and literal metaphors are probably more entangled with each other than is usually appreciated—i.e., arguably no sharp division between metaphorical and literal language necessarily exists: "At the opposite ends of a single continuum, relatively clear instances of metaphorical and literal language are fairly easy to recognize, but—except in truly dead languages—there is continual commerce between these two poles, as metaphorical concepts become more common (i.e., literal) through use and as literal concepts are used in unexpected (i.e., metaphorical) ways" (Leary, 1990b, p. 6).

13. Literalization, when it reifies entities and guides our theorizing, may cause conceptual confusion. See Barclay, 1997.

14. MacCormac contrasts "root metaphors" with "suggestive metaphors." The latter are figures of speech that have not become standardized and are used to describe novel events or strange things (MacCormac, 1976, pp. xi–xiii).

15. This discussion of metaphrands is informed by Jaynes, 1976.

16. Jaynes, 1976, pp. 55–56.

17. Jaynes, 1976, pp. 55–56.

18. Turner, 1967, pp. 28–30, 54–55.

19. Jaynes, 1976.

20. Leary, 1990b, p. 1. Significantly, the emergence of modern science from the 1600s saw a definite antimetaphorical trend (Leary, 1990b, p. 10).

21. Leary, 1990b, p. 14.

22. Smith, 1985. Cited in Leary, 1990b, p. 15.

23. Szamosi, 1986, p. 67.

Chapter Four

Unpacking the "Black Box" of Conscious Interiority

"O, what a world of unseen visions and heard silences, this insubstantial country of the mind!" exudes Julian Jaynes on the first page of his 1976 *The Origin of Consciousness in the Breakdown of the Bicameral Mind.* Jaynes, unlike many researchers, did not take our interiorized experiences for granted or explain them away as mere perceptions, biological reactions to the environment, or an unnecessary epiphenomenon of cognition (the "helpless spectator" theory). He viewed conscious interiority as a problem requiring an appreciation of a historical rupture resulting from sociopolitical complexity that occurred about 1200 BCE.

Understanding the nature of interiority is absolutely essential for coming to terms with descriptions of space and time, since they, like interiority itself, rely on spatio-visual tropes. And interiority is an example of a complex, elaborate type of meta-framing whose impact on human history has been colossal. In this chapter I itemize the contents of the package of mental abilities called interiority, but first I provide some introductory words about its nature. Let us begin with a definition. Conscious interiority is an

> analog of what is called the real world. It is built up with a vocabulary or lexical field whose terms are all metaphors or analogs of behavior in the physical world. Its reality is of the same order as mathematics. It allows us to shortcut behavioral processes and arrive at more adequate decisions. Like mathematics, it is an operator rather than a thing or repository. And it is intimately bound up with volition and decision.[1]

CONFUSION ABOUT CONSCIOUS INTERIORITY

Of all the terms used in the philosophy of mind and Psychology, "conscious" is undoubtedly one of the most confusing, so it behooves us to carefully consider the different meanings it has accrued since it first entered the English language in the early 1600s. It comes from the Latin *con* ("together") and *scire* ("to know"). Its first two meanings are alien to us: (1) "a sense difficult to define, related to a kind of animism, in which inanimate things are described as aware of human actions"; and (2) sharing knowledge with others, as in being "conscious of it one to another."[2] Eventually it acquired the meaning of "awareness," from which three other related senses developed during the 1600s. These newer meanings designate what I mean by interiority: (1) self-awareness; (2) being actively aware and reflecting; and (3) self-conscious. The latter term made an appearance in 1688, though it meant "being aware of one's actions" and sometimes denoted a great concern with one's appearance (the later additional meaning of "preoccupied with one's own personality" appeared in 1834).

During the nineteenth century, in an interesting case of borrowing a psychological experience to describe a physiological state, "consciousness" was used to designate the state of being awake (or not sleeping). Unfortunately, this usage has led to much muddled thought. To say that one is unconscious while asleep is in fact not necessarily true because when dreaming one is consciously interiorizing (i.e., introspecting one's psychoscape while sleeping).[3] Wakefulness and sleep are activities with clear neurological indicators. Presently this is not true for conscious interiority, a far more complex and subtle phenomenon. Two other interesting metaphoric elaborations in the modern usage of consciousness should be noted: (1) describing physiological states of not being in a coma; and (2) describing the mutual self-awareness of a group as if a collectivity possessed a single mind (e.g., "national consciousness," "class consciousness").

Another point of common confusion concerns the assumed location of interiority. Conscious interiority (or, the mind itself, for that matter) lacks a spatial location. Positing it in the head is an illusory act that is culturally constructed and socially taught. Historically and cross-culturally, different societies have conceptualized the location of psychological processes in very diverse places: the blood, liver, lungs, stomach, or heart. "In reality, consciousness has no location whatever except as we imagine it has." However, "volition and internal sensations" play a socially pragmatic role.[4] In other words, very practical sociopolitical reasons determine the positioning of interiority in the body, i.e., the convenience of centering control of the individual within the person (in the same manner that personal names are used by authorities to track, label, and organize populations).

Interiorized mind-space, due to its imaginary nature, has shaped a powerful folk psychology about mentation somehow literally occurring "in" the head. For this reason it leads to many mistaken assumptions. For example, the thinking of philosophers and Psychologists has labored under the following misperceptions and illusions (these confusions are treated in detail in appendix B):

- Interiority is a copy of perceptual experience.

  No: We are only conscious of a small slice of what we perceive via the senses.

- We are always experiencing conscious interiority.

  No: We are only occasionally conscious.

- Interiority reflects cognition.

  No: It is a type of cognition.

- Interiority means thinking, conceptualizing, or reasoning.

  No: Most of cognition transpires without consciousness.

- Interiority is necessary for learning.

  No: What and how we acquire knowledge occurs nonconsciously.

- Humans have always had interiorized conscious experiences.

  No: Conscious interiority, not being innately hard-wired, evolved culturally, not neurologically. As this last point is the most controversial, some elaboration is called for.

CONSCIOUS INTERIORITY AS SOCIAL ADAPTATION

History has many motors: technological innovation, religious passions, political aspirations, economic growth, demographic pressures, and natural catastrophes. Interiority is another engine of history (actually it is a collection of small motors; see "Unwrapping the Package: The Key Features of Interiority" below). To the degree that interiority is a key historical process implicated in political economic ideologies and techno-scientific inventions, these motors also describe interiority. Defining interiorization and exploring the complex historical forces that drove and strengthened its evolution are, of course, not the same. However, clearly distinguishing the meaning of a phenomenon and the conditions that gave rise to it begets the tricky "chicken-or-the-egg" problem. But we can at least approximately determine when interiority appeared on the historical stage.

As pointed out above, interiority is unnecessary for most of human behavior. If this is true, why did we as a species develop it about three thousand years ago (at least according to Julian Jaynes)? What was its purpose? From the perspective of a sociopolitical organization, interiority alleviates the need for strict hierarchical lines of command and control, which are inherently fragile. By placing a personal tool kit of command and control "inside a person's head," interiority becomes society's inner voice by proxy. Authorization based on strict hierarchical lines of command and control may be efficient for relatively small, well-circumscribed communities, but if history is any teacher, clear lines of control become less cost-effective in terms of socioeconomic capital the larger and more complex organizations become. Once authorization for immediate control of self becomes interiorized and individual-centered, an organization actually becomes stronger as its orders, directives, doctrines, admonitions, and warnings become the subjective truths of personal commitment. Interiority, then, is a sociopolitically pragmatic tool used for control in the same way assigning names to individuals or categorizing people into specialized groups for economic production is. From the individual's perspective, interiority makes the social environment easier to navigate. Before actually executing a behavior, we can "see" ourselves "in our heads" carrying out an action, thereby allowing us to shortcut actual behavioral sequences that may be time-consuming, difficult, or dangerous.

UNWRAPPING THE PACKAGE: THE KEY FEATURES OF CONSCIOUS INTERIORITY

Interiority is actually shorthand for an array of complex cognitive functions, a tool kit of contrivances that together work to form a powerful psychotechnique. How these features relate to each other (e.g., which ones came first in history) is admittedly a difficult question and beyond my present analysis. Suffice to say they are entwined, and certain features seem to elaborate or extend the meaning of others in a metaphorically rippling or cascading manner.

1. *Spatialization.* Imputing an imaginary space "inside" the individual is the most elementary feature of conscious interiority. This begins with localization (a type of spatialization) or the belief that a person's animating spirit or thinking essence (soul; what moderns refer to as "mind") is physically positioned within the person. Spatialization eventually becomes more complex as metaphoric expressions hollow out the body and lay the groundwork for psychological interiority (internal organs, e.g., "heart"). These corporeal entities are often as-

signed agency. Such spatio-interiorization generates an "introcosm" in opposition to both the immediate person (microcosm) and the external physical world (macrocosm). This interior/exterior opposition leads to dualism, an issue that still bedevils philosophers and Psychologists.

2. *Introception.* This is the ability to experience the introcosm as an inner place modeled after the perceptual, physical world. From this introscape an "I" is generated. Analogous to one's physical person, this "I" can introspect upon inner quasi-sensory experiences. Introception allows a re-experiencing of perceptual and physiological events and affects. Such mental simulations become emotions and feelings. For example, shame becomes guilt; fear, anxiety; anger, hatred; excitement, joy; disgust, contempt; and affiliation, love. The sexual drive is transmuted into carnal fantasies and eroticism.[5] Concentration, a mental act, is the analog of sensory attention, while suppression (when we stop being conscious of thoughts), "behavioral analog of repugnance, disgust, or simply turning away from annoyances in the physical world."[6]
3. *Excerption.* The inner "I" selects from the collection of possible attentions to a thing which comprises our knowledge of it. Indeed, we are never "conscious of things in their true nature, only of the excerpts we make of them." This feature of interiority is "distinct from memory. An excerpt of a thing is in consciousness the representative of the thing or event to which memories adhere, and by which we can retrieve memories." Reminiscence "is a succession of excerptions."[7] Interiority, then, does not describe a "superior unit that directs messages down to its subordinates in the brain." Rather, interiority is the "instance of selection that picks and chooses among the many options" that nonconscious processes provide for us.[8]
4. *Self-narratization.* Excerptions and mental sceneries, once unmoored from the limitations of physical reality, provide the "I" not just with a panorama of past and present events, but with imaginary "could be's." A linear temporality with the "I" as protagonist can now be constructed. Once introspectable due to spatialization, such hypotheticals form a linear temporality leading to future possibilities upon which one's "I" moves as protagonist. Interiorized cognition is a time machine, a device indispensable for long-term planning, risk assessment, and other endeavors of the modernity.
5. *Conciliation.* This describes how "a slightly ambiguous perceived object is made to conform to some previously learned schema, an automatic process sometimes called assimilation." Consilience (or compatibilization) is "doing in mind-space what narratization does in our mind-time or spatialized time. It brings things together as conscious objects just as narratization brings things together as a story."[9]

6. *Self-authorization.* One's own person ("I"), rather than external gods, ancestors, or other supernatural entities, has immediate control over one's behavior ("me"). Eventually self-authorization would bolster more democratic political organizations.
7. *Self-autonomy.* Self-authorization, as well as individual narratization, strengthens the sense of control over one's self, of personal voyage through life (self-narratization), and of destiny. Intentionality and responsibility are attributed to the "I" ("inner person") rather than divine powers, social groupings, or natural forces.
8. *Individuation.* In the same way that one's "I" is differentiated and comes to appear unique when set against the backdrop of interiorized excerptions and the journey of one's self (personal narratization), the individual's personal traits are highlighted and privileged within larger collectivities. For example, in the nineteenth century one's "character" took center stage in discussions of the relation between society and the individual. This concern intensified in the twentieth century, when highlighting individualized differences and personal uniqueness ("personality") became central to scholarly debates and everyday discussions.
9. *Self-reflexivity.* This describes the self (as a subject) observing the self (as object), or the relation between the analog "I" and a metaphor "me." This is the most difficult feature to describe. It occurs because the ability to excerpt, to "see" one's self ("I") in an interiorized place minus physical limitations, and to narratize not-yet versions of our future selves, all generate an "I" that introspects upon a "me." Such self-introspection causes a recursively regressive mirroring effect. This leads to a keenly felt existentialist perspective of a highly individuated selfness that exists in opposition to others and the world, a trend that would become salient by the nineteenth century.

I conclude this chapter with one more definition of interiority. It is

> an operation rather than a thing, a repository, or a function. It operates by way of analogy, by way of constructing an analog space with an analog 'I' that can observe that space, and move metaphorically in it. It operates on any reactivity, excerpts relevant aspects, narratizes and conciliates them together in a metaphorical space where such meanings can be manipulated like things in space. Conscious mind [interiority] is a spatial analog of the world and mental acts are analogs of bodily acts.[10]

NOTES

1. Jaynes, 1976, p. 55.
2. Williams, 1983, p. 320.

3. "Dream sleep is a state where there is no user but there are masses of illusion. . . . We cannot act with our bodies, nor can we influence our dreams with our *I*. But we are fully conscious when we dream. A user illusion without a user." Also, we "can describe a sleepwalker as a user devoid of illusions [i.e., interiority]: As he sleepwalks he acts, but the actor has no awareness of doing so" (Nørretranders, 1998, pp. 296–97).
4. Jaynes, 1976, p. 46.
5. Jaynes, 1990, p. 150.
6. Jaynes, 1991, p. 451.
7. Jaynes, 1976, pp. 61–62.
8. Nørretranders, 1998, p. 243.
9. Jaynes, 1976, pp. 61–62.
10. Jaynes, 1976, p. 65.

Part I

Space: Hollowing Out the Person

Chapter Five

Envisioning the Invisible

Spatializing the Soul

THE MAGIC OF MENTAL IMAGES

Spatialization, as the most basic feature of interiority, allows us to inhabit new lands visible only to our mind's eye. It was not until the time of the ancient Greeks that an interest in mental imagery is evident. This confirms Jaynes's argument that interiority did not exist before about 1200 BCE. Plato described such images using the metaphor of an inner artist painting on the soul, while Aristotle used the trope of making an impression on a wax tablet. During the seventeenth century philosophers often referred to mental images as a type of "idea." In the nineteenth century research on introspection was premised on the existence of mental images. More recently, interest in quasi-pictorial experiences has waxed and waned among Psychologists, while philosophers have debated their implications and very existence. The decline of behaviorism opened the door to the "cognitive revolution" in the 1950s that pursued research on inner "mental representations."

A renewed interest in the interior contents of mind has led to innovative research on mental imagery, and now there is even a journal dedicated to introspection—the *Journal of Mental Imagery*. During the 1970s attempts to conceptualize quasi-visual experiences led to the "analog–propositional" debate (probably more appropriately called the "picture–description" debate). Psychologists such as Z. W. Pylyshyn argued that mental images are not in fact "picture-like," but are better thought of as propositions or a type of cognitive language called "mentalese."[1] Others such as S. M. Kosslyn, who conducted research on the "mental rotation of images," contended that mental imagery is a nonlinguistic type of representation.[2] The analog–propositional debate has lost much of its steam, and in any case, the opposing sides seemed to be talking past each other. For example, those who claim that pictures in the head are a propositional issue have been heavily influenced by the computational metaphor model of the psyche, but Kosslyn relies on digitized bitmaps to explain how the images are generated, a decidedly "descriptive" maneuver. Furthermore, the exact nature of quasi-pictures is not clear. We might also note that though "proposition" rings seriously scientific, it was originally fancy philosophical talk for the meaning of a statement. Indeed, the picture–description debate, because its concepts are rarefied and the issues fuzzy, is reminiscent of Medieval philosophical disputations. Be that as it may, one thing is clear: The ability to introcept, as a feature of interiority, has *not* been a historical constant. Much work remains to be done on its vagaries.

In this chapter I explore and historically contextualize the "magic" of mental imagery by exploring the "soul of primitive [P]sychology," pre-Socratic thought on the soul–mind, and ideas on the soul–mind in the classical period.

The Introspection of Introcosms

Our bodies routinely perceive the surrounding environment and react accordingly with what can be called "reactivity."[3] We are not aware of most of these nonconscious reactions. The problem is that most of us confuse perception (or reactivity) with introspection. This misconception has erected the walls of a maze of misunderstanding for philosophers and Psychologists. The point is at once simple but crucial: What we ordinarily take to be "perception" (information coming via our senses) is not the same as introspection, i.e., what is "perceived" in our interiorized landscape by the mind's eye (the perception/introception distinction is sometimes referred to as "perception/apperception"). A perceptual act in physical, observable space (e.g., I see a person) is interiorized into an analogous mental event: My "I" "sees" a "person." What is introspectable is not necessarily "visible" to the mind's eye, but we usually do have a strong sense that "ideas," whether in the forms of images or imageless, exist "in" the individual, as does a self or inner person of some sort. Therefore, it is necessary to postulate a more specific kind of introspection—introception, which is a focused awareness on the "visible" contents of one's mind. Introception, then, means mental imagery, or what is sometimes called "quasi-perception." In addition to visual imagery, it may involve auditory, olfactory, haptic, or motor imagery. Introception is a type of introspection in which we experience our introcosm; more specifically, it is the interiorized counterpart of perception of the physical world. It is an extremely powerful meta-framing psycho-technique, since within our personal psychoscape, "abstract things can be treated as if they were 'things' that can be 'looked at,' 'separated out,' 'put beside,' and 'arranged' in ways that could never happen physically or in actuality."[4]

The Soul and Primitive Psychology

When we search through the historical record for concepts that resonate with modern Psychology, we often start with "soul." This is not necessarily a bad notion to begin with, but we should be aware of the differences, rather than readily stressing any misleading commonalities, with our modern ideas (this becomes clear when reading works such as A. E. Crawley's 1909 *The Idea of the Soul*). The soul has been associated with a number of notions through the ages: A vital life principle (that nourishes or sustains the body); a spiritual entity; moral discernment; and cognitive–intellectual–emotional capabilities or experiences. In many traditions, the individual often possesses multiple souls.

The soul has been an essential element in describing conscious interiority. If viewed cross-culturally, what we call soul has had two main aspects: an impersonal material life-force (usually perishable) and a personal aspect that survives death. The latter is frequently associated with individuality, agency, and intellect. In this respect, the "personal" soul seems to be what we today mean by mind. Sometimes, however, the lines were drawn a bit differently. According to Emily Michael, the soul from antiquity to early modern times performed a double role. First, it was believed to be "integrally linked to the human body as the locus or cause of its vital and cognitive activities." Second, it was "believed to be the vehicle, as separable from the human body, of personal immortality." Different aspects of this double role were "emphasized at different times and by different thinkers."[5]

Let us take a cursory examination of ancient Greek thought. For them, of course, "Psychology" as a formal academic discipline did not exist, and no disciplinary lines had yet been drawn between physiology, pathology, and cosmology. Nevertheless, early discussions of soul at least suggest an emerging, elemental psychology.

Pre-Socratic Thought on the Soul–Mind

In *Psyche and Soma*, John Wright and Paul Potter write that throughout Western history accounts of the fundamental dualism of *psyche* (soul/mind) and *soma* (body) are "so divergent in basic assumptions and goals that the question arises whether any central core of shared meaning is represented by the pair of terms at all."[6] Indeed, the soul has been distinguished from the body for a variety of purposes: religious, medical, philosophical, scientific, and psychological. To add to the confusion, the same thinker would often combine different views, sometimes holding (at least to the modern reader) anti-dualistic (monistic) views for one purpose but taking a dualistic approach for another.

Psyche and *soma* appear in the earliest extant literature of the ancient Greeks, the Homeric poems (circa 750 BCE). For the most part, in pre-Socratic Greece, *psyche* was thought of as a type of life-giving principle of the body or as something possessing the potentiality of life, concretely conceived as breath, wind, or blood (from *psychien,* "to breathe"). However, understandings of *psyche* would change over time and become bound up with sensation and purposeful movement. More abstractly, it would be related to our moral and intellectual side, or the inner person who could be self-reflective (i.e., interiorize). In sum, *psyche* would become a force representing the "teleological and integrative processes of the living organism" as well as an intellectual force constantly required for the maintenance of the body. In other contexts, *psyche* would evolve into an entity that could leave the body at death (or temporarily when fainting) and continue to exist, either in this world or the next. The lack of a clear dualism between *psyche* and *soma* is seen in how the latter could be thought of as "a kind of counter-self with desires and goals of its own" or the sensible and affective part of ourselves. On the other hand, *soma* could be more corporeal, the physical shell of a person, or an entity that is in a constant "state of corruption." It could also be conceived as a mechanical automaton.[7] At death, the individual becomes a *soma* (in this context a "corpse"), while the *psyche* ("soul"), described as breath or air, migrates to Hades where it can regain mental powers by consuming blood.[8] The dead, in fact, were the "bloodless ones"; in the Odyssey the blood of freshly slaughtered animals is fed to the dead so that they can be reanimated in order to think and speak.

In pre-Socratic times physicality and concreteness characterized what we would call psychological activity (metaphrands) which was located in bodily organs (metaphiers). Besides breath, blood, and the brain, cognition and emotion were identified with the spatial cavity of the chest: *Phrenes* may have originally meant "lungs" or "breathing" (localized in midriff); *thumos* vaguely meant "internal sensations" (sometimes localized in chest); *etor* designated heart (possibly from *etron,* belly); and *kradie* is from "heart."[9] The metaphor of visibility is clear in the term *nous* (apparently from a verb meaning "to see"). A key term for intellectual activity, *nous* was not necessarily linked to *psyche,* but was a bodily entity located in the chest.[10]

In later times the meaning of *psyche* would evolve in other directions. In the Hippocratic collection of medical writings (circa 400 BCE) *psyche* was not often mentioned. Though it was still basically considered a life principle, *psyche* was also sometimes associated with what we would call aspects of mind (sensation, dreaming, intelligence), and could be either functional or an agent of movement.[11] Interestingly, in the Hippocratic writings, *psyche*, either as life principle or as mental process, does not represent an entity with its own separate existence. In Hippocratic thought, *psyche* and *soma* were distinct, though mental events were not mere epiphenomena of bodily processes. Rather than clear-cut dualism, monism (a continuous dualism) is indicated by the belief that they both consist of the same substances. In other words, "psychic phenomena are given a somatic explanation." Mind and body are aspects of a human nature (*physis*) that embraces the "totality of bodily structures, physiological processes, and psychic events."[12] Over the centuries *psyche, nous* (mind), and *gnome* (thought) were increasingly considered to be functions, not necessarily concrete entities.

The Classical Period: Plato and Aristotle on the Soul–Mind

Plato (427–347 BCE) is significant for our purposes, and his writings tied together many themes of the cosmic split: the real versus appearance, ideas versus things, the pure versus the material, and reason and sensory perceptions. More specifically, his dichotomizing of corrupting desires (the source of trouble and immoral behavior) and an uplifting spirit (our immortal and eternal individuality) greatly influenced Christianity and Neoplatonism. This dichotomizing and meta-framing would later shape a Western philosophizing that would clearly divide the world into the grand abstractions of mind and matter.

Plato is also significant because he offers an explicit description of what for the sake of convenience we can translate as soul/mind (*psyche*) and body (*soma*) dualism.[13] However, his exposition (in *The Republic*, *Timaeus*, and *Phaedrus*) is inconsistent and sometimes contradictory. While earlier notions of thinking and the soul possessed a concrete physical spatiality, with Plato (429–347 BCE) the spatiality of such entities was becoming more abstract and refined. Plato advanced the tripartite division of the *psyche*: (1) the rational soul in the head; (2) the appetitive soul in the liver; and (3) the spirited soul or *thymos* in the chest. The last type of soul is the part of ourselves that impels us to act contrary to simple desires in order to uphold one's dignity or some other noble value.

Aristotle (385–322 BCE), Plato's student, saw a basic type of dualism in the distinction between "form" (the actual) and matter (which possesses potentiality) and viewed the soul is the actualization of potentialities. His thinking on *psyche* can be found in his *De Anima* (*On the Soul*) and *Parva Naturalia* (composed of *De memoria, De somno*, and *De somniis*).[14] Unlike Plato's plurality of souls, Aristotle's soul is unitary in the sense that different manifestations of a potentiality exist. All living substances have souls, and thinking is ultimately no more than one manifestation of a life principle. However, a hierarchy of types of the soul exists, with plants and animals as the lower forms: vegetative (or nutritive); sensitive (or sentient; in the heart); kinetic (or motive); appetitive; and the phantastic (or imaginative). However, only humans possess a noetic soul (from *nous*: "intelligence"), which is not dependent on any bodily organs and is godlike and immortal. Interestingly, Aristotle did not have much to say about *nous*, the most refined manifestation of the human soul. However, he did distinguish between an active and a receptive intellect, which suggests a distinction between will and feeling (or perhaps a distinction between motor and sensory functions).

Aristotle considered the soul an aspect of the individual, not a separate, distinct entity; the human soul is the "dynamic structure and the organizational pattern according to which, and for the purpose of which, the physical body is shaped, constituted, and internally arranged."[15] In Aristotle, we do not find a clear mind–body dualism or a distinction between what we would call physics, physiology, and Psychology. Instead, in modern terms his views may be described as psycho-physiological monism and were decidedly empiricist. For instance, he believed that the purpose of the brain is to cool the blood, and that anger results from heated blood around the heart. We remember best when our memory is "moist" and worst when it is "dry."

Terms that articulated what we understand as the mind–body split and which became central to later discussions from the classical period to later antiquity (corporeal–incorporeal, mortal–immortal, matter–form, transient–eternal, generated–ungenerated) were "not central to the more influential early Hellenistic theories of body and soul." Early Hellenistic thought is dominated by a physicalism that defies any simple dualism, and *psyche* was implicated in cognitive, moral, and bodily movement. The assumption was that all *psyche* is *soma*, but not all *soma* is *psyche*. The soul is a three-dimensional, spatially extended corporeality. An aspect of the body, it does not exist independently of our physical being. It is mortal and transient, a "fine-textured" (*leptomeres*) body. Nevertheless, *psyche* possesses special capacities, powers, faculties (*dynameis*), and susceptibilities that set it apart from the body, such as *pathe* (affections, feelings, passions). *Psyche* also is responsible for receiving sense perceptions (*aisthesis*).[16]

Such conceptualizations provide more evidence for the monistic (or continuous mind–body dualism) and is apparent in a number of thinkers, such as Epicurus of Samos (341–270 BCE) who investigated sensory experiences (and from which we derive "Epicurean"; "pleasure seeker"); the mathematician, astronomer, and geographer Erasistratus of Cyrene (circa 275–195 BCE); the physician Herophilus (335–280 BCE); and the Stoics (circa 310 BCE).[17]

I end this chapter by jumping ahead a bit toward the end of the classical period and looking at St. Augustine. This intellectual giant provided a vivid description of introspection, with his religiously motivated concern for what we would call "self-searching," "self-questioning," self-discovery," "self-description," and "self-assessment." His description is surprisingly modern in tone:[18]

When I am in this treasure house, I ask for whatever I like to be brought out to me and then some things are produced at once, some things take longer and have, as it were, to be fetched from a more remote part of the store, and some things come pouring out all together and, when in fact we want and are looking for something quite different, they thrust themselves forward as though they were saying, "Surely you must be looking for me." With the hand of my heart I brush them away from the face of my memory, until the thing that I want is discovered and brought out from its hidden place into my sight.[19]

NOTES

1. Pylyshyn, 1973, 1978, 1981.
2. Kosslyn, 1980, 1981, 1983, 1994.
3. Jaynes, 1976.
4. Jaynes, 1976, p. 60.
5. Michael, 2000, p. 147.
6. Wright and Potter, 2000, p. 1.
7. Wright and Potter, 2000, p. 7.
8. Gundert, 2000, p. 35.
9. Gundert, 2000, p. 13; see also Jaynes, 1976, pp. 257–72.
10. This a gross simplification of a much more circumspect and nuanced analysis provided by Jaynes. Relying on evidence from ancient Greek, he provides a framework for the historical evolution of linguistic terms describing psychological processes. Modern lexicons of interiority are rooted in a first phase (Objective) in which terms used for mental events originally referred to simple external things and observations outside the person. *Psyche,* perhaps originally meaning air, is a good example. The second phase (Internal) occurred when these terms became associated with processes occurring within the body, such as internal physiological sensations. Bodily organs also began to represent places where cognitive, emotional, and volitional events transpire. In the third phase (Subjective), these words actually denoted processes we would call mental: They "have moved from internal stimuli supposedly causing actions to internal spaces where metaphored actions may occur." To state it differently, objective phenomena and bodily experiences have become psychologically interiorized. The last phase (Synthetic) occurs when various terms "unite into one conscious self capable of introspection" (Jaynes, 1976, p. 260). For an application of Jaynes's framework to Japanese mind-words, see McVeigh, 1996.
11. It is not clear how many of these medical writings were composed by Hippocrates of Cos (circa 470–370 BCE).
12. Gundert, 2000, pp. 35, 29, 35.
13. Apparently for Plato *psyche* also denoted the "whole person" or the "genuine self." See T. M. Robinson's "The Defining Features of Mind–Body Dualism in the Writings of Plato," 2000. See also G. M. A. Grube's *Plato: Republic*, 1992, and Francis MacDonald Cornford's *The Republic of Plato,* 1968.
14. See *The Complete Works of Aristotle*, ed. Jonathan Barnes (2 vols.), 1984.
15. Van der Eijk, 2000, p. 63.
16. Von Staden, 2000, pp. 79, 80–81.
17. See Heinrich Von Staden, "Body, Soul, and Nerves: Epicurus, Herophilus, Erasistratus, the Stoics, and Galen," 2000.
18. Weintraub, 1978, p. 26.
19. Augustine in Warner, 1963, p. 217.

Chapter Six

Invoking Introspectable Worlds

Inner spatialization, built up from metaphors of physical places and distance, is the "most primitive aspect of consciousness." In other words, things in the physical-behavioral world that lack a "spatial quality are made to have such in consciousness."[1] This feature of conscious interiority allows one to "switch back and forth between events, processes, and facts that are widely disparate in time and space."[2] Not only do we posit mind-space inside our heads, but we assume that it exists in the heads of others.[3] This "space within the heads" of others may be thought of as an imaginary space in which dwells agency.[4] The soul has been "granted to feel spaces of time and to measure them."[5]

It is well understood that technologies shape the way we conceptualize the psyche. In the same way that our current technologies—CPUs, hard disks, computer chips (metaphiers)—are used to explain how the mind operates (metaphrands), in the ancient world daily implements were similarly used—e.g., a wax writing-tablet was utilized to describe how images are stamped in one's memory.[6]

Below I continue themes from the last chapter by examining how positing an "imaginary place" (soul–mind) within the individual was a cultural invention. Examples are given from early Christian and Thomistic thought, as well as from traditions outside the West.

INVENTING AN IMAGINARY PLACE WITHIN THE INDIVIDUAL

The ability of "picturing" things, as if some place existed within our heads in which these activities could transpire, is something we are all familiar with (if however vaguely) since we frequently rely on this form of cognition. However, these days few of us need to methodically hone our memories to any impressive degree. What is crucial for our present purpose is how the ancient art of memory relied heavily on a metaphoric notion of space. Note St. Augustine's view that memories "sink" and "slip away into some sort of distant hidden place." What "one is doing is collecting them from their dispersal. Hence the derivation of the word 'to cogitate.' For *cogo* (collect) and *cogito* (recollect) are in the same relation to each other as *ago* [set in motion] and *agito* [to vex, harry, trouble], *facio* [to make] and *factito* [to be accustomed to]."[7] Though the *ability* to metaphorically spatialize is an innate cognitive universal, actually *developing and implementing* "pictures in our heads" is a socially acquired skill that has varied historically, the same way other knowledge forms have. Indeed, the historical record appears relatively clear when the use of mental space first appeared: after the first millennium BCE.

Let us return to the classical world and look at the ancient art of memory in order to appreciate interiorized space. Cicero (106–43 BCE), the great Roman Stoic orator and statesman, related the story of Simonides of Ceos (circa 556–468 BCE) in his *De oratore*. Simonides stepped out of a banquet hall just as the roof caved in. The bodies of the victims were so mangled that their relatives could not identify them. However, Simonides was able to remember where each guest was seated just before the terrible event. He attributed this remarkable feat to memorized spatial arrangements. Cicero, crediting Simonides with the invention of the art of memory, went on to describe a mnemonic technique that relied on "places" (*loci*) and "images" (*imagines*) that are impressed upon memory.

Besides the descriptions provided by Cicero, two other texts provide discussions of ancient mnemotechnics. One is a textbook for students, the anonymous *Ad Herennium* (circa 86–82 BCE), while the other is the *Institutio Oratoria* by the Roman rhetorician Marcus Fabius Quintilianus (Quintilian) (circa 35–95). According to these works, a *locus* could be a house, corner, arch, or some other architectural feature. *Imagines* are forms, marks, or simulacra (*formae, notae, simulacra*) of what one wishes to remember. Two kinds of images exist: one for things (*res*) and the other for words (*verba*). The former pertains to an argument, notion, or thing; "words" are the language in which the subject matter is expressed.

The interiorized *loci* would come, after training, to possess a stable reality visible with a "piercing inner vision": The same set of *loci* can be used over and over for imprinting different materials. Moreover, the spatiality of *loci* should allow an individual to "start from any locus in the series and move either backwards or forwards from it." The primacy of the spatiality of *loci* is indicated by how *imagines* may fade away, but the locus can be reused (though *loci* should not be too similar to others, for such resemblance may be confusing). The *loci* are "like wax tablets which remain when what is written on them has been effaced and are ready to be written on again." Spatiality is also indicated by how the "space between the *loci* can be measured, [and] the lighting of the *loci* is allowed for"—for like the "external eye, so the inner eye of thought is less powerful when you have moved the object of sight too near or too far away."[8]

The classical sources seem to be "describing inner techniques which depend on visual impressions of almost incredible intensity," and the student was to learn a long list of rules that might seem strange to us. For example, images should not be "too brightly lighted" so as to avoid distracting glitter and dazzle; images should not be too dark, as shadows might obscure them. Indeed, these techniques seem "to belong to a world which is either impossible for us to understand or which is not being really fully explained to us." Such rules, then, "summon up a vision of a forgotten social habit" that,[9] while not completely foreign to us, points to a cognitive capability weakened by technologies of forest-destroying printing, massive information storage, and lighting-speed retrieval. This older type of memory by enhanced interiorization, however, was of crucial importance in a world whose technologies for note-taking and recording were stunningly primitive by our standards.

Here we might return to St. Augustine, a figure from the ancient Western world who is often referred to for his vivid descriptions of the "spacious palaces of memory" (*campos et lata praetoria memoriae*). St. Augustine, who was trained in the classical tradition of rhetoric, was concerned not just with what we would call sense impressions, but also with memory and other things he had learned. He wrote about "the great harbor of memory, with its secret, numberless, and indefinable recesses." Within him was a "huge court of memory": "There I have by me the sky, the earth, the sea, and all things in them which I have been able to perceive." Also in this expansive court St. Augustine encounters, in his own words, "myself; I recall myself—what I have done, when and where I did it, and in that state of mind I was at the time."[10] About a thousand years later, the Spanish Catholic philosopher Juan Luis Vives (1492–1540) wrote about how images and thoughts can be linked by association. In his *De anima et Vita* he set the precedent for the "associationists" of the seventeenth century, the "introspectionists" of the late nineteenth century, as well as those who researched "mental imagery" in the twentieth.

The idea that memory could be trained was considered essential for rhetoric and would play a role until the sixteenth century in Europe. Mnemotechnics would find its way into Renaissance magical and occult memory practices, forming complex systems of thought that linked the microcosmic (individual/soul) with the macrocosmic (nature/divinity).

We may consider the spatialization of the psyche as meaning the hollowing out of an ethereal void somewhere within the person. However, a review of the historical record illustrates that the invention of the introcosm was preceded by placing mental activity in the blood, limbs, heart, stomach, lungs, liver, the brain, or other organs. For instance, the ancient Egyptians variously placed thinking in the heart and judgment in the head or kidneys. Societies, then, borrow heavily from the body to describe psychological events. Good reasons exist for such borrowing, which are socially pragmatic, rather than inherent to the relation between body and what we conventionally call the psyche. Closer to our times the physiologist M.F.X. Bichat (1771–1802) was typical for his day in believing that the brain housed intelligence, memory, perception, and judgment while the emotions were located in the viscera (Bichat first used the word "tissue" to describe bodily materials and to distinguish between voluntary and involuntary systems). Later in the nineteenth century the notion that the brain was the "organ of the mind" was reinforced by the work of Franz Joseph Gall (1758–1828). He developed "phrenology" and believed that the parts of the brain formed separate mental faculties (though G. Spurzheim [1776–1832] coined "phrenology").

Though phrenology was essentially wrong, it did further neurology since it declared a demonstrable link between the brain and mind. Nevertheless, many of us still cling to the folk belief that mental activity somehow exists spatially "in" the brain. Though the nervous system is related to the mind in important ways, the latter does not exist "in" the brain any more than the concept "three" physically exists in three apples.

Early Christian Thought on the Soul–Mind

The early Christians distinguished the soul from body though the soul was viewed as a material substance. However, Christian theology postulated that the soul and body were a unity, not separate entities. In a time that predated the psychological hollowing out of the individual that would lead to a clear separation between person and self, individuals did not *have* bodies; they *were* bodies. Immortality did not mean immateriality of the person. For St. Paul, the "inward man" (*eso anthropos*) certainly mattered. However, a bodiless existence after Judgment Day was a foreign concept, since we will receive a "spiritual body"; for him a redeemed soul without a body was inconceivable.[11]

With St. Augustine, mind–body dualism took a new turn, i.e., the "decorporealization" and interiorization of *psyche* took one more step. We find, "for the first time, an argument for dualism that is essentially internalist. This is reasoning from what each mind can know about itself to the conclusion that minds are non-corporeal entities,"[12] i.e., the mind of each individual knows what a mind is because it knows itself.[13] This is an essentially self-reflexive act (putting a distance between the "I" and "me"). By shifting the locus of "spiritual activity from external rites and laws into the individual, Christianity brought God's infinite value into each person."[14] In other words, a newly spiritualized space, first staked out by Greek philosophers, was meta-framed and expanded into an inner kingdom where individual and Godhead could encounter each other.

The Thomistic Interpretation of the Soul–Mind

St. Thomas Aquinas (1224/25–1274), whose ideas were expressed in the *Summa Theologica* and *Summa contra Gentiles,*[15] rejected contemporary thinking about the individual that was premised on St. Augustine (who had been influenced by Plato and Neoplatonism). Instead, Aquinas adopted Aristotle's views and taught that human activity operates on three levels: (1) vegetative (growth, nutritive, reproduction); (2) sensitive (perception and motion); and (3) intellectual (reasoning and volition). Along with God and angels, humans possess reasoning and volition. There were not three different "types of souls" (vegetative, sensitive, intellectual); rather, just one soul, or a type of inherent principle that defines our nature; the human soul is the substantial aspect of the body, which it "informs." A disembodied soul is not tenable and is unnatural—i.e., the soul and the body will eventually be reunited. Though in a certain sense independent of the body, the soul requires the body for eventual resurrection—i.e., no radical disjuncture or dualism separates the soul/mind from the body; the soul is incomplete without a body.

With Aquinas, we see further meta-framing and interiorization and a clear linkage between the mental and the divine. He believed that since concepts are abstracted and reworked by nonphysical organs, whatever performs mental acts must be spiritual in nature. Aquinas sounds like an empiricist, since he believed that concepts depend on external sensory experience. In addition to the external senses of hearing, seeing, touch, taste, and smell, four "internal" senses can be enumerated: (1) common sense, which brings together "phantasms" (perceptions) produced by the external senses; (2) imagination (*imaginatio* or *phantasia*), which stores impressions and can rework them into more complex images that may not exist in actuality; (3) cogitative power (*vis cogitativa*), or a sort of base instinct allowing us to discern things as either beneficial or harmful; and (4) sense memory (*vis rememorativa*), in which ideas acquired by the cogitative power is kept.

TRADITIONS OF INTERIORITY OUTSIDE THE WESTERN TRADITION

Chinese Responses to Interiorization

How were premodern notions of interiority conceived in other civilizations? Cultural traditions associated with Chinese civilization have frequently been characterized as monistic or even antidualistic (as during the Warring States period, 481–221 BCE). However, the question of whether ancient Chinese accepted mind–body dualism is misleading. It is more appropriate to state that, as in the various traditions of the West, dualism could be either discontinuous or continuous. Also similar to earlier Western traditions, the physiological, medical, and Psychological were not separated in the modern sense. Therefore, "mental" and "physical" problems were both diagnosed holistically and would receive similar treatment (for example, acupuncture or herbal drugs). Physical, behavioral, and emotional symptoms were not distinguished. Interestingly, theorizing about what in the European tradition would eventually be called the "unconscious" was lacking. As in the West, different understandings—some popular and more "superstitious," others elite and "intellectual"—competed. More importantly, great changes in definitions of the soul and its relation to the body would transpire over time. In any case, though similarities with Western mind–body dualism can be seen, the differences are significant enough that we need to proceed with caution when making comparisons.

Let us begin with the complex array of concepts associated with yin–yang, the two primordial forces of the cosmos. Yin is the receptive, passive "female" aspect, while *yang* is the creative, active "male" aspect. From their dynamic interaction emerges the multiplicity of things, though they each contain elements of each other. How the interactions of yin and yang concern matter/mind, body/soul, and flesh/spirit are too complex to delve into for our purposes, but the relevant cosmological background at least deserves mention. Yin and yang cannot be conceived as existing in a simple opposing relationship, though their different characteristics suggest a type of cosmic split with elements of both dualistic continuity and discontinuity.[16]

Another key concept is *qi,* which is similar to the ancient Greek idea of *psyche* (from *psychein,* "to breathe," extended to mean breath, air, blood, vitality, life force). However, *qi* can also refer to a cosmic force from which all creation emanates, a sort of primal life energy from which everything, including the body, is composed. Thus it can be said that *qi* can be divided into different types or cosmic levels of material force, such as *xing* ("bodily form") and soul. Our physicality and spiritual being are of the same "cosmic stuff," though the soul comes in a more rarefied, ethereal form. However, similar to an early Greek uses of *psyche*, *qi* did not do an individual's "thinking," and it did not constitute an interiorized experience in itself.

A number of terms have been used through the centuries to describe "soul" in Chinese civilization. These were sometimes combined to denote different meanings. *Hun* meant the "heavenly aspect" of one's being, or the spirit of a person's vital force that is expressed in consciousness and intelligence. After death it was believed to gradually disperse. *Po* was the earthly aspect or the spirit of a person's physical nature as expressed in movement or bodily strength. After death it would return to the earth. *Shen* could refer to gods, spirits, or demons, but it also might denote an individual's spirit. We should also mention *gui,* which meant ghosts or demons, though when combined with *shen* (*gui-shen*), referred to ancestors. The modern Chinese translation of the English word "soul" is *ling-hun* ("*ling*" means "spiritual force").

One of the most important words for describing psychological processes in Chinese is *xin*, which literally means "heart," though it is often glossed as heart–mind. *Xin* clearly illustrates the metaphoric reliance on a bodily organ to describe interiority. Traditionally some believed that *qi* would be turned into *jing* (quintessential energy), which then would enter the *xin* and become *shen.* One's spirit, then, is a form of *qi*. This belief illustrates the continuous dualism between physicality and spirituality.

Types of dualism suggesting cosmic splits are evident in the distinction between *li* (principle) and *qi* (quasi-material life force), which was implicated in discourses about "real" or "practical learning" (*shixue*), "empiricism," and one's moral duties within the sociocosmic order. Another type of dualism is seen in the distinction between "heart–mind" (*xin*) and "things" (*wu*). A focus on individual interior experiences is clearly evident in the idealism of the Song dynasty school of Lu Xiangshan (1138–91), who stated that "space and time is in my mind, and my mind generates space and time." Later the neo-Confucian Wang Yangming (1472–1528) continued the idealist tradition with the "school of heart–mind" (*xinxue*). It might be noted that in modern Chinese "Psychology," or *xinlixue* (heart–principle–study), did not appear until around 1915.

Responses to Interiorization in India

As in the Chinese case, discontinuous mind–body dualism as it emerged in the West after the sixteenth century is not as clear in Indian civilization. However, given the complexity and sophistication of Indian psycho-philosophical thought and the uncertainty of translation, it is difficult to say with confidence what types of dualism developed. In any case, a few words are in order. Hinduism, Jainism, Buddhism, and other religious traditions formed the intellectual bedrock upon which attempts to conceptualize interiority have been built in south Asia. Many terms that might be translated as "mind" are used, such as *manas* (from *man*, "to think"). *Citta* (or *cetas*) is another relevant term, meaning heart, and is similar in use to the Chinese word *xin* ("heart"), designating the emotional, intellectual, and intentional center of the individual (*citta* can mean shine, to be bright, manifold, beautiful, tasty, sweet, spiced, and the cognate *cit* means "to perceive" or "attend to"). *Hrdaya* also means "heart" and is often used to refer to cognitive activity; both *citta* and *hrdaya* illustrate how bodily organs metaphorically indicate interiorized experiences.

Other terms include *vijñana*, from *vi,* meaning to separate or discriminate, and *jñana,* a kind of knowledge. Perceptual, intentional, or cognitive functions are also expressed by *antahkarana*, literally "inner activity" or "inner organ." "Buddhism" comes from a term used to describe interiority: *buddhi,* meaning intellect, which is from the verbal root *budh,* "to be aware" or "awake." In the same way that the *tabula rasa* has been used in Western traditions since Aristotle to describe how the outside world impresses itself upon the psyche, Indian traditions have utilized *vasana* (from the metaphor of "perfuming") to describe how any action, whether mental or physical, leaves behind "traces" or "impressions." These traces can be representational, as a memory, dream, or as a sort of psychological conditioning of the individual's mind. Agricultural tropes were also employed: seeds sown by action and experience ripen later into the "fruits" of memory and moral character. Psycho-physio-spiritual reality is also conceptualized as twelve *cakras,* or centers of mystical energy located in the body. Different types of cognition are associated with different *cakras* (incidentally, *ayuvedic* medicine resembles in many ways the humors of Western proto-medicine). In some theorizing a sense of spatiality can be found, but it is not as clear-cut as what developed in the nineteenth-century Western world, leading to the notion of the subconscious, nonconscious mental activity, and depth Psychology. Nevertheless, a type of "collective unconscious" is suggested by the term *alaya-vijñana*, which means "storehouse" or "base consciousness."

NOTES

1. Jaynes, 1976, p. 60.
2. Nørretranders, 1998, p. 126.
3. Jaynes, 1976, p. 60.
4. Whether the ability to impute mental events in others ("theory of mind") is innate or not is debatable. The capability to think about what others may be thinking about, "mindreading," is something we take for granted. We should not. For one thing, it requires years of socialization to acquire, and it is by no means clear when this capability developed in history. For instance, the cognitive ability to "take simultaneous account of the differences in the perspectives of two people" characterizes St. Augustine's great achievement in his *Confessions* (Radding, 1985, p. 39). See Andrew Whitten, ed. 1991, *Natural Theories of Mind: Evolution, Development and Simulation of Everyday Mindreading*, and Simon Baron-Cohen, 1995, *Mindblindness: An Essay of Autism and Theory of Mind.*
5. Augustine in Warner, 1963, pp. 222, 268–69.
6. See Gentner and Grudin, 1985, "The Evolution of Mental Metaphors in Psychology."
7. Augustine in Warner, 1963, pp. 217, 221–22.
8. Yates, 1966, p. 7; *Ad Herennium*, cited in Yates, 1966, p. 8.
9. Yates, 1966, pp. 4, 12, 8.
10. Augustine in Warner 1963, pp. 218, 217, 218.
11. Heckel, 2000, p. 125.
12. Matthews, 2000, p. 134.
13. See Gareth Matthews, "Internalist Reasoning in Augustine for Mind–Body Dualism," 2000.
14. Rosenthal, 1984, p. 38.

15. See *Introduction to St. Thomas Aquinas*, ed. Anton C. Pegis, 1948.

16. Another aspect of this complex cosmology was the "five phases" (*wu xing;* also called "five elements").

Chapter Seven

The Collapse of Premodern Cosmology

In premodern civilizations space as an abstract emptiness that extends into infinity in all directions hardly existed. What we call space was a collection of concrete places and locations that formed a colorful cosmology, and it was part of a religious worldview with serious political implications—e.g., even for the practical Romans, maps of the world were for state purposes and an individual might be severely punished if caught in possession of one.[1] The transition in the West from cosmography to geography was a long one of intrepid explorers, technological innovations, and a gradual desacralization of space. Indeed, until the mid-sixteenth century no word for "geography" existed in English. The first modern atlas, *The Picture of the World* (*Theatrum Orbis Terrarum*) did not appear until 1570 and was developed by Abraham Ortelius (1527–98) with the help of Mercator. This representation of the globe would mark a major spatial meta-framing.

In order to set the background for the collapse of the premodern cosmological worldview, I delineate early Medieval attempts to map the globe as well as the inner geography of the soul. This breakdown ushered in new "spaces"—macroscopic, microscopic, and introscopic—and would lead to the "measurement revolution," a significant meta-framing in intellectual abilities that heralded scientific modernity.

MAPPING THE EARTH AND SOUL

The flexible nature of space makes its mapping much more contingent and unstable than we usually assume. Consider that for a thousand years during the European Middle Ages, the findings of Ptolemy (90–168; Claudius Ptolemaius of Alexandria) had been forgotten, and the accumulated knowledge of local explorers was ignored. Indeed, a common word for geography was lacking. Ptolemy's *Almagest* (written around 150) contains geographical information and the needed trigonometry to explain a geocentric spherical cosmos. This work greatly impacted the Arabic world in the ninth century and Western astronomers and geographers several centuries later, though the worldview it afforded was not systematically challenged until the early 1600s.

During the Medieval era, Christian maps of the world were biblically inspired. These were called "wheel" or "T-O maps" because they were typically circular and divided by a T-shaped flow of water. At the top of the map was Asia in the east, while below on the left of the vertical part of the "T" was Africa in the south. Europe was on the other side of the "T" in the north. Jerusalem constituted the sacred center of the known world, and the terrible nations of Gog and Magog were in the north, ever threatening Christendom. The mythical realm of Prester John, who ruled over a type of earthly Christian utopia, was often represented, and up until the late 1500s, maps still commonly indicated his Kingdom. The Garden of Eden or Terrestrial Paradise was usually imagined somewhere in the east, and as in classical times, Hell was believed to be an actual place one could visit, if one happened to be interested. Throughout the centuries it was commonly described as a subterranean land of horrors or as occupying the lower part of the globe.

Legendary realms and regions of the earth—Atlantis, Shangri-la, El Dorado come to mind—have played a mythical role, blurring the boundary between fact and fable. For example, many searched for the Great South Land or what was called the Southern Continent, which was represented on seventeenth-century maps. In the late eighteenth century the great explorer Captain James Cook was tasked with disproving its existence.

By the Medieval period geography, then, had become "a ragbag filled with odds and ends of knowledge and pseudo-knowledge, of Biblical dogma, travelers' tales, philosophers' speculations, and mythical imaginings."[2] In the premodern era of map-making, bodies of water and land masses were often represented within a stylized figure of some sort (e.g., a tree), without consideration for actual relations; the point of the representation was allegorical correspondences, not geographical accuracy.[3]

Spatiality and temporality are bound up with each other in subtle ways, and meta-framing in one entails changes in the other. For example, the cartographer Gerardus Mercator (1512–94), who devised a way to depict the spherical earth on a flat surface for use by seafarers, also provided a universal time line.[4] By the early modern period, the "itch to use space and time had broken out." Once time and space were coordinated with movement, they could be "contracted or expanded," meaning that the "conquest of space and time had begun."[5] Not coincidentally, it was not until the sixteenth century that the notion of acceleration was formulated.

It is worth pointing out that while map-making in the Western world stagnated in the Middle Ages, China made stunning advances. Indeed, "the last two millennia in China display a galaxy of cartographic talent." While Christian map-makers worked hard to make their maps reflect scriptural texts, Chinese cosmographers were achieving accomplishments impressive by modern standards.[6] Moreover, evidence exists that a rather sophisticated and accurate model of the globe was brought to China from the Islamic world in the 1200s.

Visions of the Medieval *Mentalité*

Let us delineate the macrocosm of Western Medieval times. The world was conceived as stacked in layers. Although there were many variations on how best to conceive the cosmos, an earthly terrestrial realm, where things were composed of earth, air, fire, and water, was a common starting place. The heavenly celestial realm was subdivided into different levels, in which things were made of the fifth essence, "ether."[7] Planets moved through a transparent, ethereal material, and the "true order of space was Heaven even as the true order of time was Eternity."[8] Beyond the celestial realm, or we might say "outside" physical spatiality, was the Empyrean of God, an ultra-transcendent realm were space and time did not operate. Though discontinuous in nature, these different spaces were part of one supercosmos in which angels, spirits, and human souls floated about. Physical and spiritual cosmos were interwoven, with the spiritual realm being the "primary reality" and the terrestrial realm a less pure imitation of the spiritual. In the European Medieval vision literature, then, the theological and physical were different worlds, but holistically formed a cosmic unity. Thus, the factual and the fantastic intermingled. This accounts for the prevalence of strange beliefs. For example, journeys to the other world were considered real during the Medieval period, even if such travels occurred in dreams. And one could enter the "mouth of Hell" in certain places, such as Lough Derg in Ireland.

In the Medieval mind, spatiality and temporality formed two relatively independent systems. For example, in Gothic painting space and time were "successive and unrelated," so that the "immediate and the eternal, the near and the far, were confused." Indeed, the ordering of time was distorted by "the jumble of events presented and by the impossibility of distinguishing hearsay from observation and fact from conjecture." Medieval artists would think of nothing of comingling different times within the same representation. For example, the events of Christ's life might be introduced within a contemporary Italian city "without the slightest feeling that the passage of time has a made a difference." Each scene in the portrayal was equally near to the observer. It was only later when individuals related events to a coordinated, unidirectional frame of time or a homogeneous spatiality extending in all directions that "being out of time or being untrue to time became disconcerting."[9] We moderns usually think of pictures (especially photographs) as depictions of what existed at a single point on linear time, a frozen frame of what was happening in a "knife-edge instant." However, Medieval representations might include "several clearly differentiated 'nows'" (i.e., moments in time or different historical periods). The viewer could look backward and forward in time simultaneously. An analogy borrowed from spatiality might help here: As "we pass a cubical building, we perceive it not in a durationless instant in which we can never see more than two walls, but *as we move,* and so we can, sometimes, catch sight of three walls in one 'now.'" Though somehow empty, space for us is a "three-dimensional vacancy around the subject or among the subjects."[10] We do not deny the existence of space because we cannot view it; though airy and hardly visible, it possesses a "thingness" that we take for granted. But for the Medieval individual space was what it contained; it did not exist unless it was concretized by filler of some sort. A vacuum was impossible. By the same token, time was cluttered with events, episodes, and incidents. It was what happened, not an empty tunnel through which we pass, taking part in occurrences.

A purely physicalist paradigm was "quite alien to the Medieval mindset," while we moderns "simply cannot imagine a place being 'real' unless it has a mathematically precise location in physical space."[11] And note that since it would take centuries before mathematics divorced itself from mysticism, magic and numbers would be interconnected for centuries. It was assumed that being possessed of a certain neatness, orderliness, and predictability, numbers themselves were reflections of more a perfect, transcendental realm. Numerals were not just quantitative values with practical functions; they were infused with qualitative properties that linked the microcosmos and macrocosmos together. As Crosby notes, the modern marriage of mathematics and measurement was not as inevitable as we might assume. More abstract representations of numerical functions would greatly enhance the meta-framing of mathematics; one obstacle to the marriage was that most mathematical operations, except for counting and simple arithmetic, were expressed in written form without signs for plus, minus, square root, etc.[12] Moreover, the modern microscopic and macroscopic scales simply did not exist in the Medieval cosmic vision.

Daniel J. Boorstin cautions us not to scoff at the Medieval worldview too much. After all, it was an intellectually coherent system which described the "heavens precisely as they looked and fitted the observations and calculations made with the naked eye." Moreover, "it actually performed some functions of a scientific explanation. For it fitted the available facts, was a reasonably satisfactory device for prediction, and harmonized with the accepted view of the rest of nature."[13]

The Collapse of Medieval Cosmology

The implosion of Medieval cosmology was long and complex, involving many developments. It is a specific example of a widespread crumbling of the premodern worldview. As I discuss below, these developments involved the introduction of perspective, the discovery and exploration of the macroscopic and microscopic worlds, the refinement of cartography, and most importantly for our purposes, the interiorizing of the person. It is worthwhile to draw out the linkages between spatiality and Psychology, since as we shall see, the discovery of new physical spaces was part of the process of finding new territories within the person; and as older notions of the soul were left behind, the imaginary horizons of the introcosm expanded.

Perspective as an Interiorization of the Macroscape

Between the fourteenth and the seventeenth centuries a major change in conceptions of space occurred. This transformation was related to the discovery of the laws of perspective. It is well known that Gothic artists had "almost no sense of depth or solidity," and lacked an interest in the illusion of three dimensions. Though the ancient Greeks used perspective to create the illusion of depth, they apparently did not know or apply its mathematical laws.[14] "Seeing geometrically" is not as natural as we may think.[15] Though linear perspective was used to convey the religious import of paintings in the 1400s, it was eventually secularized and helped lay the foundations for the scientific revolution.[16] Perspective transformed symbolic-laden arrangements of objects into visual relations. In turn the visual became a coordination of quantitative relations. In this new world of abstracted relations, "size meant not human or divine importance, but distance." Artists used an "abstract checkerboard" to represent reality and establish "point for point correspondence between the picture and the image."[17] Designers began to simulate physical space by using faux architectural details to create a "powerful illusion of actual architecture that has the effect of blurring the boundary between the virtual space of the image and the physical space of the chapel."[18] The architect, engineer, and inventor Filippo Brunelleschi (1377–1446) wrote on mathematical perspective, and Leon Battista Alberti (1404–72), the brilliant Renaissance humanist polymath who theorized on perspective, wrote *Della Pittura* (known in Latin as *De Pictura*) in which he utilized classical ideas on optics to theorize on perspective. The artist Piero della Francesca (ca. 1416–92) also published several mathematical treatises on the topic. Tommaso Masaccio (1401–28) was representative of the newfound aesthetic sensitivity to perspective. Through the use of *chiaroscuro* (Italian for "light dark"), he conveyed the reflection of light from surfaces that added a sense of three-dimensional depth. These factors in artistic representation indicated a "newfound interest" in the natural world that radically challenged the Medieval mind-set of physical and spiritual space. Perspective, like other developments that had to with a changing worldview impacted by an emerging appreciation for measurement, concerned calculation, quantity, mathematical preciseness, and geometrical consistency—all things we take for granted.

Perspective brought not only depth into the picture but "distance into the mind"; or if phrased differently, perspective, as a type of meta-framing, highlighted a key feature of interiority: the spatialization of the psyche into an introcosm. Perspective means placing the observer into the depicted environment while positioning the mind's eye within the individual in such a way that it "looks out" from the person. It granted the mind's eye a more multidimensional and textured introscape to observe relative to actual perception. Moreover, what the "painters demonstrated in their application of perspective the cartographers established in the same century in their new maps."[19] This is a case of the microscape (small space) informing the macroscape (large space). The rules of perspective highlight the difference between perception (what one's senses detect or "know") and introception (what one's "I" "sees"). Perspective is not just about quantifying reality; it is also an attempt to grant the introception of reality an inherent value.

By the 1400s and 1500s, artistic representation and new forms of metaframed visuality greatly improved scientific and engineering illustrations. Around this time certain pictorial conventions—cutaway views, perspective sections, transparent views—were invented. Engineers, architects, anatomists, and botanists utilized these to show the reader what words cannot describe clearly.[20]

The Discovery of Astronomical and Macroscopic Space

The emergence of the astronomical scale was part and parcel of the toppling of the Medieval conception of the cosmos. The story has been told many times, but here it is worth mentioning how the development of the heliocentric theory by Nicolaus Copernicus (1473–1543) overthrew Ptolemaic cosmology. In *Concerning the Revolutions of the Heavenly Spheres* (1543) Copernicus described the cosmos in fundamentally mathematical terms. It is easy to forget how revolutionary Copernicus's contribution was. He not only defied the church-ordained geocentric theory (wisely he did not publish his findings until he was on his deathbed), but he heralded the coming of modernity with its view of the infinity of space and the abstractness of time. Our world would be transformed from a cosmos of animistic, pulsating forces to a universe of cold, precise mechanical laws.

Johannes Kepler (1571–1630) was another important contributor to the emergence of astronomical spatiality. An astronomer and astrologer for the

Holy Roman Emperor Rudolf II, he studied the laws of planetary movement and optics. In his *Mysterium Cosmographicum* (*The Sacred Mystery of the Cosmos*, 1596) he attempted an ultimately unsatisfactory geometrical explanation of planetary trajectories (at that time only five planets were thought to exist). However, his *Harmonie Mundi* (*Harmony of the Universe,* 1611), which also contained elements of mathematical mysticism, provided a brilliant arithmetical account of planetary motion. His astronomy, being informed by the aesthetics of music, was as much art as science. For centuries, it was believed that each planet possessed its own "spirit" or "intelligence" (*mens*) which guided it through the heavens. Kepler used the term *animae motrices* ("moving spirits"), but he eventually replaced "soul" or "spirit" (*anima*) with "force" (*vis*). This reconceptualization helped usher in the demise of animism.

Galileo Galilei (1546–1642), in addition to his path-breaking work in astronomy and motion, made numerous other contributions, such as his development of the scientific method. His ideas on gravitation formed the basis for Newton's work. He improved the telescope (which had been invented in 1608 in the Netherlands), and may have invented the compound microscope in 1610. Though Galileo wrote mostly on astronomy and physics, he intuited that these realms of knowledge were linked to other areas, as apparent in his investigation of perception and sensation in his *The Starry Messenger* (1623). In *Dialogue on the Two Chief Systems of the World* (1632) he compared the Ptolemaic and Copernican systems. This landed him in trouble, and he had to recant his defense of the Copernican worldview after being tried by the Inquisition.

The Discovery of Microscopic Space

The discovery of microscopic vistas, as in the case of macroscopic horizons, was made possible by technological innovations. The inventor of the microscope is uncertain, though some believe it was Zacharias Jansen (1588–1631?), a spectacle-maker and counterfeiter of Spanish copper coins. We do know the name of the coiner of the new device: the physician–naturalist John Faber (1574–1629). The mathematician and scientist Robert Hooke (1635–1703) explored and detailed the newly discovered microscape in *Micrographia* (1665). In the same work he also wrote about light and color, as well as other things. Antoni van Leeuwenhoek (1632–1723), a fabric merchant and draper who spoke only Dutch and had no formal scholarly degrees, explored the world of microbiology, discovering, among other things, bacteria. The Italian lens-maker Eustachio Divini (1610–85) used a microscope to describe a world of invisible, tiny animals. The founder of "microscopic anatomy" and the new science of microscopy (first used in English in 1664) was Marcello Malpighi (1628–94).[21]

Remarkably, even after the invention of the microscope, not a few resisted the exploration of new microspaces. Why did it take some time before the microscapes revealed by the microscope were accepted as real?

Arguably the answer is in how sight itself was conceived. Vision required illumination, and during the Medieval era light was not a mere physical phenomenon but a manifestation of the sacred; to study its physical properties was "like inquiring into the chemistry of the Eucharist." And like vision, light was regarded as somehow mystical. Light and seeing were not to be toyed with, and some held a strong distrust of optical devices. For centuries it was believed that objects did not emit physical impressions registered by the eye; rather, vision was a process of the eye itself, an "active" organ. Though eyeglasses were used in Europe during the late Middle Ages, mirrors, prisms, and lenses were devices for distorting the nature of light and associated with trickery. Sight was a God-given faculty, and anything positioned between the eyes and the observable world could easily form "optical illusions" (*deceptiones visus*).[22]

The Western world was indeed in the "dark" ages when it came to optics, since elsewhere, more sophisticated approaches toward optics had been developed. The brilliant Arab philosopher and polymath Al-Kindi (Abu-Yusuf Ya'qub ibn-Ishaq; 803–873), who lived in what is now Iraq, believed that rays moved from illuminated objects to the eye. And Alhazen (Ibn al-Haytham; 965–1039) carried out sophisticated experiments on dazzle and the persistence of images on the retina. He demonstrated that vision was not a product of the eye (though he understood it as a piece of optical machinery); rather, the external world played a defining role in visual perception.

Interiorizing the Body

Beginning in the late Medieval period dissection and other advances brought a visible realism to conceptualizations of the human condition and literally interiorized the body. Up until approximately the 1400s, medicine was more a branch of theology, not a science in itself, and it was meant to confirm cosmological truths rather than empirically delineate the world of nature. The works of the ancient Greek physician Galen (circa 130–200) dominated medical knowledge for fifteen hundred years. Galen, who treated gladiators, the Emperor Marcus Aurelius (121–180), and the latter's son, Commodus (161–192), relied on the dissection and vivisection of pigs and monkeys, since in his time human dissection was forbidden.

Objective investigations were a major blow to the Galenic tradition, and a momentous shift from a concern with mysterious qualities to measurable quantities transpired. Humors and vital spirits would no longer be unquestionably relied upon to account for human psycho-physiological behavior. For instance, Andreas Vesalius (1514–64), whose gruesome body-snatching and detached, meta-framed observations of the quartering of criminals must have shocked his contemporaries, wrote *De Humanis Corporis Fabrica Libri Septem* (*The Fabric of the Human Body in Seven Books*, 1543). Vesalius, whose work concerned the psycho-physiological question of whether the heart was the seat of the soul (metaphrand), was interested in *fabrica* (metaphier), a word that generated paraphiers and paraphrands, suggesting both the body's "structure" (from the analogy of "fabric") and its "workings" (from the analogy of the French *fabrique* or the German *Fabrik* for "factory"). Santorio Santorio (1561–1636), who wrote *Ars de Statica Medicina* (*On Statistical Medicine*, 1614), developed what he called "static medicine" (from the Latin *staticus* and the Greek term for the art of weighing). Mathematically oriented, he realized the importance of measurement for medicine. And William Harvey (1578–1657) in his *On the Motion of the Heart and of Blood in Animals* (*De Motu Cordis et Sanguinis in Animalibus,* 1628), showed how the heart functioned and that blood, rather than *pneuma* ("animal spirits"), was the key component of the circulatory system.[23]

THE MEASUREMENT REVOLUTION AS META-FRAMING

Since recorded time people have measured things, of course, but according to Alfred W. Crosby, sometime around 1250 to 1350 a signature feature of what was to be the macro–microscopic worldview began to emerge.[24] This feature is something so much a part of our everyday world that we hardly question it: measurement. What was distinctive about the thirteenth century was how broad the range of what could be quantified had become as well as the concern with preciseness (including more careful quantification of that very insubstantial stuff, time). This "devotion to breaking down things and energies and practices and perceptions into uniform parts and counting them" is a powerful instance of meta-framing. By around the 1500s Westerners were treating the cosmos in terms of uniformed measurements, such as lines, squares, and circles. But this transition to symmetrical forms was not just one of representation. As Crosby points out, music staffs, platoons, ledger columns, and planetary orbits began to be conceptualized as geometrical abstractions. How individuals experienced visuality itself had undergone a transition: painters began conceiving of scenes as "geometrically precise visual cones or pyramids focused on the observing eye."[25]

NOTES

1. Boorstin, 1985, p. 269.
2. Boorstin, 1985, p. 100.
3. Mumford, 1934, p. 19.
4. Boorstin, 1985, p. 599.

5. Mumford, 1934, p. 22.
6. Boorstin, 1985, pp. 111, 113.
7. After all, "different levels of reality require different spatial domains: *A multiform reality demands a mutliform conception of space*" (Wertheim, 1999, p. 128, emphasis in original).
8. Mumford, 1934, p. 20.
9. Mumford, 1934, pp. 342, 19.
10. Crosby, 1997, pp. 166, 168 (emphasis in original), 170–71.
11. Wertheim, 1999, pp. 55, 71.
12. Crosby, 1997, p. 14.
13. Boorstin, 1985, p. 295.
14. Wertheim, 1999, p. 79.
15. Crosby, 1997, p. 176.
16. Edgerton, 2009, *The Mirror, the Window, and the Telescope: How Renaissance Linear Perspective Changed Our Vision of the Universe.*
17. Mumford, 1934, p. 20.
18. Wertheim, 1999, p. 79.
19. Mumford, 1934, pp. 20, 21.
20. Crosby, 1997, p. 231–32.
21. Boorstin, 1985, p. 376.
22. Boorstin, 1985, pp. 313, 328.
23. Boorstin, 1985, pp. 359, 362, 371–73.
24. Crosby, 1997, p. 18.
25. Crosby, 1997, pp. 11, 10–11.

Part II

Psyche: The Origins of Scientific Psychology

Chapter Eight

The Foundations of the Modern Study of Mind

An "independent [P]sychology is absent, both in thought as well as in fact, from philosophical systems of antiquity."[1] Indeed, though the word "psychology" was used in English, French, and German before the nineteenth century, the term was primarily implicated in philosophical, theological, moral, medical, and political discourses. Until the early or mid-nineteenth century it was not unambiguously related to what we would call "psychological." As an autonomous discipline, "Psychology" certainly did not exist, either theoretically or institutionally, and by implication, the introcosm upon which individuals introspected was not problematized as a target of scientific scrutiny. Debates during the late Middle Ages and Renaissance were informed by an Aristotelian worldview that divided the introcosm into the *animus* and the *anima.* The former was the rational, intellective, immortal, immaterial, divine soul; or what we might call the mind as opposed to the body. The latter was the irrational, organic, mortal, material, animal soul; or what was regarded as the bodily, sensitive, or vegetative soul. The *animus* was associated with *mens,* a concept we might refer to as introspectable "mind" in which reasoning occurs, while the *anima* was the life-principle animating the body.[2] The soul and body were believed to form a unity, not a union. Clear-cut Cartesian dualism would come later.

In this chapter I investigate the intellectual antecedents to the modern study of the psyche and the emergence of proto-Psychology.

PRE- AND PROTO-PSYCHOLOGICAL: INTELLECTUAL ANTECEDENTS

When the earliest references to "Psychology" first appeared is a topic of debate, and deciding what earlier authors meant by allied terms is somewhat of a challenge.[3] In any case, "Psychology" was apparently first used as a Latin word (*psychologia*) by the Serbo-Croatian humanist Marko Marulic (1450–1524) in 1520[4] (*psyche* originally meant "breath" and later "soul" in ancient Greek).[5] *Psychologia* had little to do with the "science of the human mind." It was a subdivision of pneumatology, a term commonly used until the late 1700s that meant what we would consider the psychological (for example, by Leibniz). Pneumatology, or the science of spiritual or incorporeal stuff, was subdivided into three orders of spiritual substances: God, angels, and the human soul. Psychology dealt with the latter entity.[6] Philipp Melanchton (1497–1560), an administrator who implemented Luther's reforms in the universities of Central Europe, used the Latin "*psychologia*" in his writings about theology, the senses, and control of the passions. The German scholastic philosopher Rudolphus Goelenius (or Rudolf Goeckel; 1547–1628) used the term in 1590 in *Psychologia, hoc est, de hominis perfectione* (*Psychology or On the Improvement of Man*). He was interested in the "nature of the human spirit or logical soul" and focused on its "immortal intellective" aspects. Otto Casmann, a student of Goelenius, wrote *Psychologia, anthropologica, sive animae humanae doctrina* (1594) and divided *anthropologia* into *psychologia* (the human mind) and *somatologia* (the human body). The encyclopediast Johann Heinrich Alsted (1588–1638) divided pneumatology into *empsychologia* (the study of the soul as a life-giving force animating the body) and *psychologia* (the disembodied soul).

Theology, philosophy, physiology, physics, pedagogy, mental hygiene, and political economics: these are all possible routes that, if retraced, might lead us back to the origins of institutionalized Psychology.[7] It all depends on to what degree a researcher reads into the writings of a predecessor the meanings associated with the psychological. Writings of the late eighteenth century on physiology, sex, marriage, children, madness, and the rise of literature (particularly the novel) do suggest proto-Psychological themes. Nevertheless, the seventeenth century is "devoid of texts comparable in aims or methods to modern [P]sychology." However, by the 1770s a psychologized atmosphere existed that would foster the eventual development of Psychology.[8] And we should keep in mind the numerous figures who have undoubtedly been overlooked in histories of Psychology. For example, Robert Whytt (1711–66), a figure of the Scottish Enlightenment, developed some basic assumptions behind scientific Psychology, such as the doctrine of stimulus and unconscious sensations. He coined "stimulus" and experimented on the spinal reflexes of vertebrates.[9]

With the aforementioned stated, the tendency to project into the writings of earlier thinkers our own assumptions must be guarded against. It is not necessarily unwarranted to conclude that "what they meant was something we would call psychological." However, it is intellectually perilous to then conclude that they were consciously moving forward to and laying the foundations for modern research Psychology. Edward S. Reed notes that in the eighteenth and early nineteenth centuries we do not actually know what constituted Psychology. After all, doctors, academics, preachers, writers, and quacks wrote about what we now call the psychological.[10]

A cast of esteemed characters, such as René Descartes, John Locke, Thomas Hobbes, Gottfried Leibniz, George Berkeley, and David Hume, is usually presented as a troupe of the "first Psychologists" or "early Psychologists." Such designations are highly problematic. These personages are being forced to perform a role on the stage of history that they would undoubtedly find strange. It is important to remember that key personages in the history of Psychology were not, of course, in the mold of modern research Psychologists. For example, Descartes (1596–1650) framed the modern discussion of the mind–body debate by severing the nonmaterial mind/soul (located in the pineal gland) from the material body. While the latter possessed extension and motion, the former was nonphysical. He simultaneously carved out a space for scientific investigation (our corporeality) while reserving a supernatural realm beyond the laws of physics, a point worth making since this bifurcation of reality is in many ways still with us. But the Cartesian legacy is arguably more clearly evident in the revolutionizing of mathematics and provision of a philosophical foundation for the natural sciences. Descartes also made contributions to optics and physiology. His famous statement—*Cogito ergo sum* ("I think, therefore I am")—was more of an epistemological claim about how ways of knowing can be approached.

Even in the early and mid-nineteenth century, notable personages often credited with contributing to the rise of Psychology were not, by the standards of the twentieth century, Psychologists. For example, Sir Charles Bell (1774–1842), Pierre Flourens (1794–1867), Johannes P. Müller (1801–58), and Ernst Heinrich Weber (1795–1878) were physicians and physiologists. James Mill (1773–1836) was a historian and diplomat; John Stuart Mill (1806–73) was a philosopher and political economist; Gustav Fechner (1801–87) was a physicist and philosopher; Hermann von Helmholtz (1821–94) was a physiologist and physicist; and Franz Brentano (1838–1917) was a priest and philosopher.

Graham Richards argues that three late-eighteenth-century intellectual movements are "directly ancestral" to Psychology: phrenology, physiognomy, and mesmerism. He notes that the former two both employed empirical methods, were applied in a practical way much like Psychology would be (e.g., counseling and evaluation of others' character). Moreover, they "rapidly became identified with reform and optimism about the meliorative value of a Science of Man." Most significantly, "neither was completely assimilable to either philosophy or medicine."[11] In other words, phrenology and physiognomy suggested a space within the individual not reducible to the endeavors of logical reasoning or physiological investigation.

Despite any caution required in viewing the aforementioned key figures as advancing Psychology, all helped transform the introcosm into an introscopic realm amenable to scientific scrutiny. Driven by developments and quantum leaps in technicalization, these intellectual pioneers assisted in the interiorization of the individual, giving more explicit attention to an individuated, isolated, invisible space within each person. Psychology as a scholarly field is a meta-framing built up from spatio-visual metaphors.

THE EMERGENCE OF PROTO-PSYCHOLOGY

How can we discuss Psychology if our predecessors lacked such a concept? For my purposes, "interiority" will cover what we moderns refer to as conscious subjective mind. Interiority, of course, was not the word they used, but enough intellectual resonances exist between what I denote by this term and the recorded subjective experiences of our predecessors to justify its use (such as a crucial aspect of one's personal identity dwelling in, through, or around the body, as well as mental imagery). However, the contents and contours of interiority have developed considerably through the ages, and these vagaries must be recognized. After all, in the pre-Psychological ancient world interiority was characterized by what we might call religio-philosophical or medical concerns. Thus, rather than speaking of the history of Psychology in premodern times, we should speak of religious interiority (discussions about the soul and spirit). In the late seventeenth century, philosophical approaches to interiority emerged (speculations about the "inner sense") that for the most part were not amenable to scientific or experimental approaches. Categories of thought that we consider separate formed seamless bodies of knowledge, and works that appear "modern" were actually often grounded in thinking that is quite alien to us.

For example, consider Robert Burton's (1577–1640) *Anatomy of Melancholy* (1621). Though more a depository of literature-inspired learning than a medical work, this book deals with what we call matters Psychological. However, its advocacy of sweating, purging, bloodletting, and vomiting to restore one's balance of "humors" illustrates how its worldview was informed by a psycho-cosmology that had more in common with the ancient Greeks (from where the idea of humors come) than our own.

Until the mid-nineteenth century Psychology as we think of it today was a branch of philosophy, religion, or physiology. However, a new space embedded within the individual was gradually recognized as not being necessarily of a spiritual nature. In any case, taking a spiritual stance toward interiority was for some unscientific. And philosophy, which pursued abstruse problems that bordered on mysticism, did not seem to clarify the issues. Physiological approaches could not adequately capture the experience of inner experience either. Individuals interested in interiority used physiology and physics to lay the groundwork for experimental Psychology (at first called "physiological Psychology"). Perhaps we can speak of medical interiority (e.g., studies of brain processes and the nervous system) and sensory Psychological interiority (e.g., studies of perception).

Over the *longue durée* we can see how increasing interiorization paralleled the intellectual trajectory of modern experimental Psychology. It was during the nineteenth century in Europe that different intellectual currents formed a confluence from which a modern Psychology, premised on interiority, developed. Many saw a linkage between the brain and the mind (that was more of a novel idea than we might at first assume, since traditionally mind could be found anywhere in the body). This medico-psychologization of the person is evident in the work of a number of pioneers, such as Pierre Florens (1794–1867) who studied the diversification of the function in the central nervous system, Luigi Rolando (1770–1831) who studied the anatomy and pathology of the brain, and Paul Broca (1824–80) who localized the brain's speech center ("Broca's area"). Around the same time investigators pursued the conduction of nervous stimulation and sensation. Early Psychology grew out of physiology and was originally called "physiological Psychology" and then "experimental Psychology." Moreover, many researchers considered their work on the mind to be experiments in physics, so that the realm of psychologicality was not yet independent from a physicalist view of experience.

Interestingly, the development of sensory physiological Psychology followed a certain trajectory: first vision, then hearing, and then a bit later, touch. That objects, due to their inherent nature, must be experienced visuospatially, the most basic medium for our sensory experiences, probably accounts for why sight attracted such interest. Recall that in the Medieval period the nature of light was the topic of much proto-scientific theorizing, and later many great thinkers, such as Newton's *Optiks* (1704), studied light and vision.

Eventually researchers used "introspectionism" in order to describe interiority, and by the end of the nineteenth century they spoke of experimental Psychology. A different terminology would be used in the nineteenth century to describe what we refer to as psychiatry or clinical Psychology: "medical Psychology" or "mental science"; a psychiatrist was called an "alienist," "medical Psychologist," or "mental physician."

Studies of introspection would make certain features of conscious interiority central to the Psychological endeavor. Then, the assault of behaviorism on the invisible side of the human experience would deny interiority as a topic worthy of serious Psychological research. Despite the attempts of behaviorism to squelch work on interiority, by the latter half of the twentieth-century research Psychologists spoke of "mental imagery" when referring to the introspectable properties of interiority. The orthodoxy and excessiveness of behaviorism would give way in the 1950s to the "mental" or "cognitive revolution" which readmitted modern psychological interiority onto the research agenda.

NOTES

1. Lapointe, 1970, p. 640.
2. Vidal, 1993.
3. For instance, in *De trinitate* (*On the Trinity*), St. Augustine used both the Latin feminine word for soul (*anima*) as well as the masculine word (*animus*); the latter term seems to suggest "a rational *anima*—that is, a rational soul, or mind (*mens*)" (Matthews, 2000, p. 137).
4. Krstic, 1964.
5. See Brožek, 1999, "From '*Psichiologia*' to '*Psychologia*': A Graphically Documented Archival Study across Three Centuries."
6. Vidal, 1993; Lapointe, 1970.
7. Note that histories of Psychology neglect sixteenth-century theologians who "invented the word, and barely mention or treat in the most 'presentist' manner, such later writers of 'Psychologies' as Wolff and Bonnet" (Vidal, 1993, p. 90).
8. Richards, 1992, pp. 231, 15, 246.
9. Reed, 1997, pp. 231, 230.
10. Reed, 1997, pp. x, xi.
11. Richards, 1992, pp. 249, 250, 265.

Chapter Nine

The Great Cosmic Split

Dualism

The external/internal and visible/invisible metaphors of human experience may be related to what we can term great "cosmic splits" that characterize all cosmologies. These dissections of reality have formatted our modes of thinking and realms of social existence. In this chapter, after introducing different types of dualisms generated by cosmic splits, I trace the roots of modern Psychology to a "new spatiality." Then the intellectual attempts to mend the cosmic split are discussed, as well as the various accounts of mind–body dichotomies.

WEAK AND STRONG DUALISMS

Cosmic splits—or we may simply say, dualisms—come in two varieties. The first type, which characterizes ancient and premodern thinking, regards reality as continuous (loosely a type of monism)—i.e., absent is a sharp break between people and gods, the corporeal and spiritual, nature and humankind, the flesh and soul. Rather, qualities mundane and spiritual shade into each other, the profane and sacred interpenetrate. We might call this "weak" dualism. Though in Medieval Europe one's more immediate surroundings were on occasion suddenly and noticeably infiltrated by the other world, it was assumed that in an important sense existence was diffused with divinity. Living in a world permeated by the numinous, demonic possessions and usurpations were to be expected, as were the more welcome divine visitations and inspirations.

The second type of dualism, which describes the modern perspective from around the seventeenth century, considers reality as coming in compartmentalized, discontinuous segments. We might say that the world is partitioned into various "subrealities," reflecting sharp rifts in existence ("strong" dualism). The view that varying degrees of mundane-ness and spirituality constitute our existence became less acceptable. Though belief in another world has by no means disappeared, the burden of proof gradually migrated from experimental and empirical approaches to the religiously minded. The shift from "and/both" to a "versus/or" type of dualistic distinction is certainly not complete, and within many of us several paradigms typically coexist.

Whether a continuous or a discontinuous dualism, it is difficult to find a tradition of thought that has not been shaped by the cosmic split. It is, indeed, very difficult for us to escape from mind–body dualism. For example, consider "psychosomatic medicine." Far from "reflecting a monist or holist view of the human being, [it] instead reaffirms a physical/psychologist dualism in the very moment that it attempts to elucidate its components' interactions." Viewed from this perspective, it is a late product of the new dualism whose beginnings were in the eighteenth century.[1] And the religious thinking that we still rely on attests to the staying power of ancient sentiments: substance/spirit, worldly/holy, humankind/God, mundane/mystical. Epistemological ruptures beget the key philosophical debates: mind/matter, idea/real, subject/object, physical/metaphysical, consciousness/world. The ghosts of older dualisms shape medical knowledge, as seen in debates about treatment: a humanistic, person-centered, and "interpretative" Psychology (mind) versus a more physiological, biomedical, or drug-based psychiatry (body).

Cosmic splits may seem to rupture existence, but actually they often link—or mirror—different levels of reality. For instance, St. Augustine believed that the Trinity (macro-level) was reflected in the three powers of the soul at the microlevel: memory, understanding, and will. The emergence of the social sciences is premised on disciplinary divisions and themes that reflect the split: psychology/sociology, subjectivity/objectivity, self/others, belief/behavior, cultural/material. Crucial discussions in political thought are motivated by the problem of "inner freedom" versus "external control": agency/structure, individualism/groupism, individual/society. The placement within the individual of a private conscience, something inviolate and sacred, guaranteed by public authorities (at least in theory) during the seventeenth century, or the secular world/religion division, reveals the spiritualization of the political and the politicization of the spiritual. Our daily language, loaded with all sorts of metaphrands, metaphiers, paraphrands, and paraphiers, bespeaks the cosmic split that still haunts us: eye/mind's eye, matter/meaning, manners/morals, material/immaterial, senses/thinking, reality/dream, tangible/intangible, objective/subjective, visible/invisible, concreteness/abstractness, stuff/symbols, thingness/information.

Let us consider the conceptual topography of one social science: Modern scientific Psychology, whose very development, beginning in the seventeenth century, was driven by a fundamental fracture forcibly asserted by René Descartes (1596–1650). Though he was not the only thinker to concern himself with the perceived differences between conception/perception and introspection/vision, Descartes described what he took to be a fault line between mind/soul and body. The Cartesian worldview is significant because it clearly and explicitly declares a discontinuous duality between an "inner" mind and an "outer" body, a split that has become for us a psycho–philosophical *idée fixe.* The origins of mind–body dualism are grounded in the spatial metaphors of mind: Cognition, emotions, and volition (metaphrand) are described by a location that can be observed (metaphier). The latter spins off the paraphier of the mind-as-container ("inside" vs. "outside" person), which bolsters the paraphrand of "hidden, invisible thoughts vs. revealed, visible behavior."

In the Western tradition "introspection is actually a historically recent invention. Its roots may go back to certain tendencies in Protestant theology that advocated a methodical self-examination of one's conscience." The thinking of Descartes and Locke was only a "first step," and for them introspection "remained a potential object of technical scrutiny." The "inner world of the isolated individual mind" is not as unproblematic as many of us assume.[2] Anyone who has taken the time to ponder the nature of interiority will appreciate the difficulties. Consider the words of Edwin Boring, a historian of psychology: "The persistent problem of introspection is the question as to whether in a perception, for example, the perception itself is also a knowledge of itself, or whether we have also to 'perceive' the perception to know that we have it, thus presumably establishing an infinite regress."[3]

A NEW SPATIALITY: THE ROOTS OF MODERN PSYCHOLOGY

Descartes cannot take all the credit for the great cosmic split identified with his name, but his clear prose delineates a milestone in the history of interiorization.[4] The rupture of reality he theorized about actually manifested itself in a number of smaller cracks that nevertheless have shaped the modern worldview (e.g., body/mind, soul/body, visible physicality/invisible thought). Descartes completely rewrote our definition of and approach to interiority. It was as if he described a massive earthquake that rearranged the entire geography of the seeable physical and unseeable spiritual realms.

A cold mechanical and lucid mathematical logic characterizes Descartes's ideas on human nature (that also informed his other great scientific works). For him the cosmic split that explained so much was the difference between two radically different substances: Measurable, material, divisible *res exten-*

sa ("extended thing") and immeasurable, immaterial, indivisible *res cogitans* ("thinking thing"). Our bodies are extended substance, while our minds are thinking substance. Prior to Descartes, many thinkers had regarded mental activity as a part of the soul. Descartes took a revolutionary step by narrowing the meaning of the soul and clearly identifying it with the mind. It was a meta-framing maneuver of momentous intellectual impact. For Descartes, the only function of the soul is to think (e.g., what we might associate with interiority). There are no vegetative or sensitive souls. The mind is not a part or aspect of the soul; rather, the "soul is the mind in its entirety."[5] *Mens* (mind) is *anima* (soul). It is not multiple and lacks parts or aspects. The soul is indivisible. A century or two later similar understandings of the person would develop in the political sphere (inviolability of conscience, one vote for one person, etc.). Bodily functions that were previously imbued with animistic soulness were relegated to the realm of biology and squarely placed on the extended-substance side of the cosmic divide. Bodies are machines that do not think. Being imbued with physicality and movement no longer meant that a creature was ensouled. In this way, animals were de-souled and became mere automata.

Unlike soul-less animals, humans had one foot in extended substance and the other in the thinking-substance side of existence. Thinking substance and extended substance met in the pineal gland. However, the pineal gland was merely the point where the mind interacted with the body—it was not the seat of the soul; indeed, arguably the entire body is the seat of the soul, since the activity of the soul can affect the whole human organism. More precisely, the pineal gland was where the mind determined the flow of animal spirits throughout the body. Being immaterial, the soul did not occupy space, but it did require a spatial point in the brain to have its effect (Descartes did not identify the brain with the mind). The fact that the pineal gland is not duplicated (i.e., not divisible), as is the case with other parts of the brain, earned its special role of bridging the cosmic divide. Some have described Descartes's dualism as mind–body parallelism (i.e., no interaction between the two). This is not completely true, since he obviously believed in a type of interactionism between the body and the soul. Moreover, "animal spirits" mediated between the soul and the body (note that Gottfried Wilhelm Leibnitz (1646–1716) advocated psycho-physical parallelism and believed that mind and body did not interact; they are not causally related and follow different paths).

Descartes's contribution to the future development of research Psychology is complex and rich, and his theorizing represents a passage from an older animism to a universe governed by mechanistic principles associated with the modern worldview. For instance, he believed that animal spirits (though now corporeal and despiritualized) originate in the heart and flow through the pipes or hollow tubes of the nerves; the muscles were engines while the brain acted as the executive office for physical control. This mechanistic approach, the interaction of mind and body, the brain as the primary locus of mind, and localization of the mind throughout the entire body (and "yet the specific localizations within the brain")[6] are all traceable to the Cartesian cosmic split. Moreover, Descartes's "excision of the soul from nature and his exclusion of its study from positive science—that is, physics—were deeply revolutionary, to an extent we have still not fully grasped."[7] Descartes did not completely discount the role of sensory impressions. However, by emphasizing the innate nature of reasoning, volition, and a sense of what we might call "selfness," he highlighted interiority—e.g., self-reflexivity and self-autonomy of one's own thinking. Also, though his own religious belief was questioned during his time, by moving the soul from scientific scrutiny to the realm of unprovable spirituality, Descartes ironically ensured that an aspect of the interior experience would retain a religious element.

ATTEMPTS TO MEND THE COSMIC SPLIT

In "Renaissance Theories of Body, Soul, and Mind" Emily Michael points out that from classical times to the early modern period the soul performed a double role in relation to the body: (1) it was the source of the body's vital and cognitive activities; and (2) it was separable from the human body and acted as the medium of personal immortality. During the Renaissance, debates about the soul progressed from agreement on the harmony of these two roles to an obsession with their conflict.[8] In other words, concomitant with the increasing spatialization of the psyche, the dualism between soul/mind and body was becoming sharper. Why?

During the Renaissance the Thomistic interpretation of Aristotle's *De Anima* and *Parva Naturalia* was very influential for shaping definitions of the soul and its relation to the body.[9] Aristotle's belief that the soul has different powers (vegetative, sensitive, appetitive, intellective, etc.) influenced theories about how the individual possessed more than one soul. Some philosophers even claimed that the soul, since they identified it with sensory and purely intellectual functions unrelated to a religious agenda, was mortal. The Church was understandably suspicious of such thinkers. Thus, in the early sixteenth century new non-Thomistic interpretations were required to demonstrate the immortality of the soul.

The intellectual processes behind these aforementioned changes are more complex than I am portraying them here, but it is clear that novel categorizations of the soul–mind–body totality were emerging. The groundwork for early modern accounts of the mind as something ultimately beyond the reach of the material world was developing. While some argued that the sensitive organic soul was a "medium" between body and mind, others saw a crucial distinction between a *mortal* corporeal sensitive soul and an *immortal* incorporeal mind.[10] The result was that many began to postulate that the body (and its related vitalistic or organic aspects) cannot think; only a soul/mind could do that.

In the Western world, then, the Renaissance played a crucial role in the transition from Medieval accounts to more meta-framed, modern understandings of the body, soul, and mind. In addition to theological disputes, debates about the soul/mind–body relationship were probably motivated by a struggle to come to terms with an emerging interiorization driven by a changing social environment. The increased number of social roles performed by the individual, challenges to the political economic authorities, and the destabilizing impact of rapidly circulating wealth meta-framed cognition as it enhanced self-reflexive interiority, so that it became advantageous for individuals to apprehend their own apprehension—"thinking can turn upon itself and think about this thinking."[11] This self-reflexive ability destabilized traditional definitions of the soul as mere organic processes grounded in bodily substantiality or as a heaven-bound entity. As new functions of interiority forced their way onto the scene, some concluded that the thinking soul—or the mind—did not require corporeality to function. Reflection seemed to be a power of the immaterial soul, and this spiritual incorporeality would lead to the "mind" in the modern sense.

EXPLAINING MIND–BODY INTERACTIONS

At this point, we can briefly delineate three approaches to the mind–body mystery that had emerged in clear form by the nineteenth century. The first is the mentalist (or idealist) stance: some sort of mental stuff is real, while the external body is an epiphenomenon. The second stance, physicalist or materialist, merely reverses mentalism: the external body is real, while the mind is an epiphenomenon. The third approach is dualist: both mind and body are phenomena in their own right; the challenge is to explain how they relate to each other. The dualist approach can be subdivided into: (1) mind-as-prime mover, i.e., the relation is one way and the mind moves the body; (2) body-as-prime mover, i.e., the relation is one way and the body moves the mind; (3) parallel theories (or psychophysical correlation): mind and body cannot effect each other; they are helpless spectators in relation to each other; (4) interactionist (or psychophysical interaction): body and mind mutually affect each other. The most influential type of dualism, in terms of shaping the West's postmedieval philosophy, has been Cartesian (which with some qualification is interactionist): body and mind are different types of substances, i.e., one is a physical and extended (body), while the other is spiritual and nonextended substance (mind). Our spiritual substance (or mental attributes) is independent of our physicality, but nevertheless it interacts with the body.

The different dualist accounts may not solve the mind–body problem. However, though mentalists, physicalists, and other antidualists (such as monist, who view body and mind as different aspects of the same entity/substance) may claim they have a better take on the mind–body puzzle, in fact their theorizing is still fundamentally premised on dualism.

In more recent times some have dealt with the mind–body problem by pushing it to the margins of respectable inquiry. For instance, behaviorism, which in many respects dominated American Psychology for the first half of the twentieth century, simply ignored the issue, declaring that mental events, too closely associated with soul stuff, were not suited to scientific investigation. In *The Concept of Mind* (1949), the British philosopher Gilbert Ryle (1900–76) criticized the Cartesian worldview, arguing that Descartes and other philosophers had allowed language to mislead them by assigning one type of word to a realm of description for which it was inappropriate. Though our use of metaphoric language has convinced us of the necessity of postulating an invisible "mental world" different from observable behavior, nonspatial events such as thinking do not require another "second world" to transpire; the body does not need a "ghost in the machine." The work of Gilbert Ryle is highly sophisticated, clarifying as it does the misleading use of language when discussing mental events. However, it does not explain why some historical periods privileged introspectable experiences. Indeed, the various approaches do not explain why the mind–body problem—upon which modern Psychology as a discipline is founded—arose in the first place. My answer is that due to interiorization, the experience of an "inner person" ensconced within a body "looking out" at the physical world became stronger (such an experience is, after all, something learned). This visuospatialization was driven by a growing focus on the individual (individuation) over the centuries (figure 9.1).

ANOTHER DUALISM: *PHYSIQUE* AND *MORAL*

I conclude this chapter with another type of dualism. By the late eighteenth century, a new division of human nature had emerged: *homme physique* versus *homme moral*. This cosmic split became "as commonplace as the pair body/soul a century earlier" and was represented in such works as the two-volume *Rapports du physique et du moral de l'homme* (*On the Relations between the Physical and Moral Aspects of Man*, 1802), written by Pierre Jean Georges Cabanis (1757–1808).[12] The *physique/moral* distinction was originally applied to the realms of aesthetics and morals, but eventually it acquired a role in medicine.

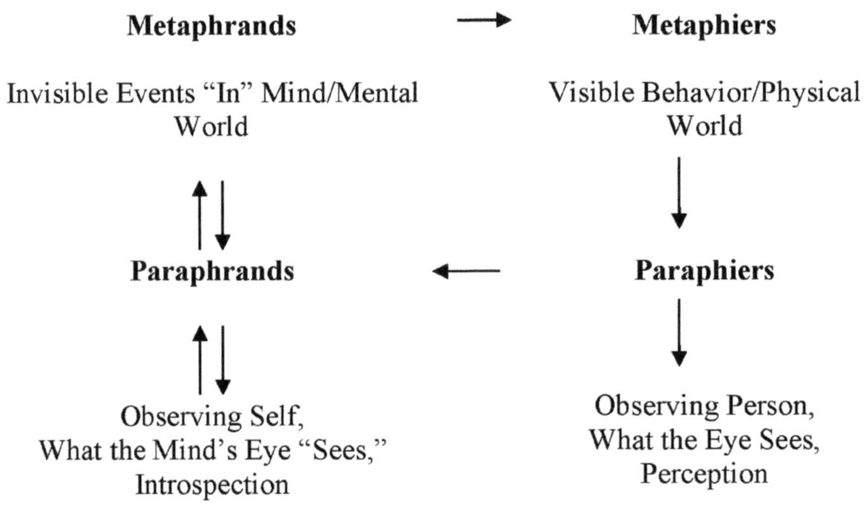

Figure 9.1. Visual Metaphors of Mind and Mind–Body Dualism.

Physique meant the "real, existing in nature" with mathematically certainty. *Moral* designated that which "concerns morals, conduct of life" and was associated with probability. It might involve what we would call interiority, such as sensing, reasoning, imaging, or remembering. "*Moral* is introduced, it seems, when the necessity is felt to designate a realm of experience which is opposed not only to that of physical nature, but also to that of the soul." Physical man is the peripheral nervous system, while moral man is the central nervous system.[13] According to Cabanis, the other name for moral man is "interior man, that is, the cerebral organ."[14]

Why did this type of dualism emerge? In "*Physique* and *Moral*" François Azouvi writes that Cartesian dualism "dug out a chasm between physical and moral man which subsequent philosophy and medicine have striven to fill up." A new idiom was needed to conceptualize "something different from what was understood by the Cartesian pair, body and soul. *Physique* served to designate something bodily which cannot be reduced to the merely mechanical, *moral* something mental which cannot be reduced to simple consciousness." After the Cartesian revolution the soul became identified with the mind, so a new term was needed to designate the "opposite of material reality." Today we conventionally refer to this realm as "psychological." Moral, then, would be the "ancestor of *psychique* [English: "psychological"], referring to something non-material, which is nevertheless not pure thought."[15]

The problem of how to relate the intellectual/moral realm to its organic/vital causes left many questions unanswered. According to Azouvi, aspects of the soma migrated to the psyche, while aspects of the psyche migrated to the soma. However, how these migrations actually occurred remained a mystery.[16] The philosopher Maine de Biran (1766–1824), whose work foreshadowed psychoanalysis, psychopharmacology, educational Psychology, and phenomenology, would seriously entertain these problems. For him, the will, with all its moralistic implications, was realized by an active introspection. However, he believed that physiological approaches would have to accompany any investigation into our "inner senses." In any case, it was interiority that has given rise to this and other theories of *homo duplex*, i.e., a being in which the instinctual demands and desires of the human animal compete with social norms and pressures to obey ethical rules and moral dictates.

NOTES

1. Azouvi, 2000, p. 279.
2. Danziger, 1994, pp. 218, 21.
3. Boring, 1929, p. 163.
4. See his *Discourse on Method,* 2006 (1637); *Meditations on First Philosophy,* 1996 (1642); *The Passions of the Soul,* 1989 (1649); and *The World,* 2006 (written in 1633 but published posthumously).
5. Voss, 2000, p. 175.
6. Boring, 1929, p. 164.
7. Voss, 2000, p. 176.
8. Michael, 2000, p. 147.
9. Michael, 2000, pp. 147–48.
10. Michael, 2000, p. 165.
11. Michael, 2000, pp. 147, 161.
12. Azouvi, 2000, p. 268.
13. Azouvi, 2000, pp. 270, 271.
14. Cited in Azouvi, 2000, p. 272.
15. Azouvi, 2000, pp. 276, 270.
16. Azouvi, 2000, p. 278.

Chapter Ten

Reactions to the Cartesian Split

The intellectual tremor caused by the Cartesian cosmic rupture saw responses from the best thinkers of the time. This cosmic crack, however, did not sit well with everyone; not all believed that the human soul could be circumscribed by the straight lines of geometry or the unambiguities of an ordered, mathematicalized worldview. Descartes's sharp dualism did not take into account a whole array of human behavior that does not fall cleanly under the body or soul/mind side of the great divide. These behaviors, such as spontaneous, instinctive, involuntary, or automatic bodily processes, do not require conscious attention. As it became clear that nonconscious psychic activity did not register in the introcosm, thinkers were forced to reconsider their ideas on the relation between body and mind. In this chapter I look at responses to the Descartesian clear-cut dualism of mind and body by examining British empiricism and German nativism.

A COSMIC SPLIT: EXTERNALIST VERSUS INTERNALIST NOTIONS OF KNOWLEDGE

For any number of reasons, different traditions prefer to emphasize one end of the cosmic split. For example, in seventeenth-century Europe attempts to understand interiority and its relation to the world had gone in opposite directions. The Anglophone world emphasized how interiority was the product of what we would call socialization, and for the English the real world of pragmatic action and common sense were guides to appreciating human nature. In other words, experience was empirically formed, by a sort of association of bits of information, the mind (a "mental chemistry"). Across the English Channel on the continent, interior experience was considered the result of a limited number of innate operations. These nativist processes were regarded as manifestations of rationality, an essential attribute of the human condition that allows us to discern the relationship between our existence and knowledge. The French made a fetish out of logic and sought an orderly relationship between what we think and the world (though eventually empiricism would impact French philosophy). In Germany, nativism would eventually develop into an idealism mixed with Teutonic mysticism and grand theory-building.

Both sides of any debate offer insights, of course, but the empiricist-versus-rationalist argument itself evidences increasing interiorization. In other words, the more internal definitions of mind became (internalism), the more it became obvious that human experience was also deeply implicated in a world exterior to the psyche (externalism). The empiricist-versus-rationalist debate also indicates the parameters of psychic diversity—i.e., the human mind is not whatever we decide to make it, but is configured by both external (social) forces and internal (innate) features.

APPROACHES TO CHANGING CONCEPTIONS OF PSYCHE

Some attempted to maintain premodern animistic notions in their accounts of human nature. For example, the medical theorist Georg Ernst Stahl (1659–1734) advocated an antimechanicalist theory of physiology and pathology. He proposed an animism—though not one grounded in a metaphysics—in which "bodily structures are formed, preserved, repaired, and functionally adapted by the action of an immaterial principle, the soul." The soul is active and immaterial, the "efficient and final cause of the vital and organic motions, as well as the conscious affectations of the human subject." He opposed a mechanist view that denied the existence of organic forces, as well as any absolute distinction between soul and body characteristic of Cartesian dualism.[1] Some phenomena, however, are indeed explicable using mechanical principles (that in turn can be expressed mathematically). They lack "immanent dynamic purposes: Specific appetites, attractions, sympathies, etc." However, "a true organism" is the "site of an autonomous instrumental relationship," that is, the body is a means through which the soul operates.[2]

In "Psyche, Soma, and the Vitalist Philosophy of Medicine" Roselyne Rey writes that Cartesian dualism "had opened up a radical discontinuity between the activity of the rational soul, on the one hand, and the mechanism of the body, on the other, making possible a study of the body alone."[3] In particular, a group of physicians in the second half of the eighteenth century rejected any simple dualism (or we might say they were more receptive to a continuous dualism rather than a sharp break between mind and body). These vitalists also rejected the view of animists who, "in restoring the soul's power over the body in organic functions" (e.g., digestion and secretions) had "given too much to the soul, and taken too much from the body." Phrased differently, the animists had created a "purely passive bodily machine" that lacked a vitalist principle. This principle was *sensibilité* (sensibility), a general property of all living things. More than a mere intermediary between the body and mind, it constituted its own psycho-physiological entity. The vitalists believed in the unity of the physical and the moral, which were aspects of the individual that reciprocally influenced each other. In other words, mental and physical states were considered parts of the same continuum.[4] The immaterial and immortal soul, now unnecessary for organic functions, was separated from the body. For their part, the so-called vitalists, by not equating sensibility with thought, opened the door to an investigation of anomalous behavior such as somnambulism, dissociative disorders, the mysteries of sleep, effects of habit, temperament, and other phenomena not easily categorized as either psychological or physiological. Vitalists "called into question the simplicity and oneness of psychic activity." Though they did not arrive at the notion of the "unconscious" as some claim,[5] their recognition that simple mind–body dualism is limited was a meta-framing step toward the scientific scrutiny of interiority.

In "Substance versus Function Dualism in Eighteenth-Century Medicine" John P. Wright explains how by the 1700s two types of dualism had emerged. These represent two sides of the cosmic split. The first dualism was one of "substance." The soul–body distinction was absolute since these were substances of "entirely different and incompatible natures" or essences: "material body substance" and a "non-material soul substance." Substance dualism, which was Cartesian though it could be traced back to antiquity, was in part sustained by religious sentiment that required an otherworldly substance. In this sense substance dualism opposed crude materialism.[6] The second dualism, upon which psychosomatic medicine was premised, was "functional." In this type the mind or soul had "thought functions," while the body had "life functions."

Despite attempts to maintain a religious element to interiority, conceptions of the mind were breaking free from two historically stubborn notions: (1) an individual's spiritual center (immortal soul); and (2) vital life processes (e.g., vegetative soul). The liberating of an abstract, incorporeal, and meta-framed "thinking thing" from concrete substances and religious doctrine would lead to modern Psychological definitions of mind as an object of study in its own right. It would take, however, another two or three centuries before modern scientific Psychology would clearly delineate the mind. Nevertheless, the mystery of our introspectable abilities, surrounded by vagueness and misunderstandings, still bedevils modern Psychology and philosophy.

BRITISH EMPIRICISM

Locke as Proto-Psychologist

We hear so many times how profound the contributions of an intellectual giant such as John Locke (1632–1704) were that we end up taking their wisdom for granted. But like all of us, he was imperfect. For instance, Locke's religious tolerance did not extend to Catholics, and his plans for combating pauperism were shockingly cold-hearted, even by the standards of the day. In any case, he stands as a founding father of the Enlightenment. In addition to his modern-sounding political theories, his recognition—along with Hobbes—that an understanding of the interiorized individual was necessary to appreciate new political economic arrangements is a shining insight.

That experiences somehow impress themselves on the mind is not new with Locke and can be found in Aristotle. However, in *An Essay Concerning Human Understanding* (1690)[7] Locke makes empirical experience central to his proto-Psychological theorizing. Though other thinkers of his time had discussed "ideas," for Locke these "objects of the mind" took on an importance and specificity until then not appreciated. Ideas could be sensations, memories, concepts, or mental images. They had two sources: sensation (observing external objects) and reflection, an "internal sense" in which the operations of the psyche are observed. The combination of simple ideas generated more complex ideas. Through comparison we obtain ideas of relations, and by abstracting these we form more general ideas (ideas about ideas; an instance of meta-framing). Locke's belief that the mind could be analyzed into ideas just as matter could be broken down into smaller parts was influenced by his friend, Robert Boyle (1627–91), the founder of modern chemistry. This notion of "mental chemistry" would play a salient role as "associationism" ("combination" is the term Locke actually used).[8] What is significant is how Lockean empiricism opposed Cartesian rationalism (the innateness of ideas). Locke is also significant for his comments on the self-reflexive function of interiority and individuation, or how personal identity is a matter of the unity of conscious experience.

Hobbes and Materialist Psychology

In *Leviathan* (1651) Thomas Hobbes (1588–1679) systematically justified and built a theory of sovereignty by beginning with the individual mind that senses the impressions that shape it. Since Leviathan was a political entity constructed by humans for humans, the very nature of the latter needs to be understood. Hobbes was actually reacting to and acknowledging how an aspect of interiority, or individual intentionality, was implicated in new political arrangements. These greatly differed from Medieval institutions (but note that the secular thrust of Hobbes's work must be qualified, since half of *Leviathan* treats religion). Rather than God-ordained and organic, for Hobbes social institutions were guided by human will and purposefully designed. Such intentionality required a new understanding of what motivated people. Influenced by the new worldview being shaped by the astronomical work of Kepler and Galileo, Hobbes saw individuals behaving like so much matter in motion. Hobbes also corresponded with Francis Bacon, the proto-scientist and advocate of inductive reasoning. Rational agreements, contracts, and a political system designed along geometrical principles were the order of the day. Hobbes's materialist Psychology is crude and mechanistic: individuals are driven by the fear of death and desire for power, and mental events are reducible to the motion of matter in the body. Nevertheless, the empiricism of his Psychology rings modern: mental events accompany or are causally dependent on physicality.

Hume and the Laws of the Mind

The empiricism of Locke was developed by David Hume (1711–76), who wrote *Enquiry Concerning Human Understanding* (1748; the earlier and longer version is called *A Treatise of Human Nature: An Attempt to Introduce the Experimental [Newtonian] Method of Reasoning into Moral Subjects*). This luminary of the Scottish Enlightenment and philosopher from Edinburgh, who also penned many ideas on economics, politics, religion, and history, made great use of introspecting upon the contents of his own interiority for insights into the psyche (though certainly he was not the first to do so). From these inner observations he derived some of his theories. Hume contended that when he "looked" into his own mind, he detected no "impression" of the self. Therefore, there was no "idea of self," or no impression corresponding to an "I." Instead, he witnessed only a mere "bundle" of sensations but no unifying idea. The items enumerated within his introcosm were nonphysical, and while the body remained, the mind was decidedly ephemeral.

Hume's contribution to what would become Psychology was his insistence that there existed a graded distinction between "impressions" (sensations or perceptions; but note Hume used "perception" to mean both impressions and ideas) and "ideas" (reflections or what we might call introception; note that Locke used "idea" more broadly). For Hume, impressions, which cause ideas, have more of a direct impact on the mind, while ideas are faint copies of impressions. Impressions seemed to mean "sensate" (feeling, passions, emotions, what is immediately felt), whereas ideas are "non-sensate" (thinking, reasoning, "faint images"). Ideas can be divided into memory or imagination. This impression–versus–idea distinction no doubt sharpened dualism and enhanced belief in individual interiority.

Hume accepted Locke's notion of the compounding of simple ideas into more complex ones, but he made his predecessor's associationism more explicit. The influence of Newton on Hume's thinking is clear, since he considered association, like gravity, a "kind of attraction" operating within the introcosm of the mind. In the same way astronomers searched the macrocosm for the laws of the heavens, Hume sought the laws of mental causation. Association was a "uniting principle" of the mind. It took three forms: resemblance; contiguity (in time and place); and cause and effect.

The Humean skepticism toward a unitary self is open to a number of interpretations. I suggest that in the idiom of the time he was challenging the assumption that somewhere within the individual there dwells a homunculus (literally, "manikin" or "little man"). He intuitively understood that the self is a function of our minds, not an executive ego or "inner master" of our person. It is only one performer on the private stage of our inner theater and reflects certain aspects of our sociopsychological being; it is not a scriptwriter or director of mental activities.

Berkeley: The Introcosm as Primary

If for Descartes the problem of the cosmic split was about how mind related to matter, then for Locke it was how matter generated mind. For George Berkeley (1685–1753), however, the problem was how mind generated matter.[9] Berkeley, the Bishop of Cloyne in County Cork in Ireland and known for his *A Treatise Concerning the Principles of Human Knowledge* (1710), is famous for subjective idealism, or the affirmation of mind as the ultimate reality. We might say that for Berkeley the interiorized spaces of the psyche have been turned inside out to such a degree that they occupy the whole of reality. This is a radical type of introspection: perception is reality or "to be perceived is to be" (*esse et percipi*). The world exists because God thinks it does. Or if Descartes was convinced that "I think, therefore I am," Berkeley believed that "I think, therefore I am in the mind of God." It is as if he took one feature of interiority—the spatialization of mind—and built a philosophy upon it. Despite criticisms of extreme solipsism, Berkeley made important contributions to the Psychology of "meaning" and visual space-perception, as evident in his *An Essay towards a New Theory of Vision* (1709).

Hartley's Psychophysical Parallelism

Empiricism and associationism continued and found explicit fruition in the work of the physician and philosopher David Hartley (1705–57). Indeed, due to Hartley's systemizing of contemporary thought in his *Observations on Man* (1749), associationism became a doctrine. His attempt to grasp reality, like many theories at the time, was very much Newtonian: vibrations in matter caused corresponding vibrations in the nerves. Hartley did not accept the then fashionable notion that "animal spirits" flowed through hollow tubes; correctly, he believed the nerves were solid. The "vibratory actions" of the nerves, like strings in an instrument, resulted in smaller, subtler vibrations in the brain ("vibratiuncles," the physiological counterpart of ideas) which then formed thoughts. The mind, then, operated by sound-like vibrations that resonate with other ideas. This vibrational, associational Psychology makes Hartley a primitive physiological Psychologist. For Hartley, mind is not yet an abstracted, interiorized space within the brain. Though ideas only exist in the brain, the "seat of the mind" is the brain, spinal cord, and nerves, all of which concern sensation and motion. Unlike the interactionist dualism of Descartes, Hartley's dualism is best described as psychophysical parallelism.[10]

John Stuart Mill and Spencer and the Birth of Psychology

Associationism would play a crucial role in theories of mind in the eighteenth century. The political economist and historian James Mill (1773–1836) is well known for his *History of British India* (1818). However, he also wrote *Analysis of the Phenomena of the Human Mind* (1829) in which associationism played a defining role. Mill advocated a rather minimalist type of mental mechanics and a simplistic dualism that echoes the mind–body disconnect: the mind has only ideas (conceptions) and sensations (perceptions). Mill's son, the philosopher and political economist John Stuart Mill (1806–73), continued this associationism. His thoughts on Psychology are found in *Logic* (1843) and *Examinations of Sir William Hamilton's Philosophy* (1865). He modified his father's rather mechanical mental chemistry. Instead of coming together in aggregates, ideas fused into completely new ones.

In the ambitious scholarly hurricane that was Herbert Spencer (1820–1903) we find a confident propagandist for the dominant faith of the nineteenth century: evolutionary progress. The certainty of science was replacing the comforts of religion, and belief in a purposeful universe in which humankind had reached the pinnacle of cosmic development was very attractive. This undoubtedly explains the contemporary popularity, which cannot be underestimated, of Spencer's ideas on biology, Psychology, sociology, and ethics. Though we do not now take his ideas on social evolution seriously (which are not the same as Darwinian natural evolution), for our purposes his *Principles of Psychology* (1855) is important because in it he discussed associationism. For Spencer, an unfolding hierarchy from simple sensations to more complex states defined an evolutionary associationism.

GERMANIC NATIVISM

Leibnitz and "Little Minds"

Rationalism and Teutonic mysticism were married in the genius of Gottfried Wilhelm Leibnitz. Lawyer, diplomat, discover of the calculus (along with Newton), and all-round polymath, Leibnitz, in order to solve the mind–body problem, thought up an elaborate philosophy. The result was an example of an intellectually coherent system that, taken as a whole, is strikingly strange and has convinced few; nevertheless, it possesses particular insights and useful ideas.

He conceived the world as an infinite plurality of independent "monads" (from *monas,* Greek for "unity"). These might be considered as "little minds" that are indestructible, uncreated, immaterial, and immutable. They constitute reality as we experience it, though actually what we experience is the monads perceiving each other. Individually monads lack perception, but when in the aggregate, they form a type of mentality. Monads, though their essence does not change, are in a constant state of development and may obtain higher degrees of perception or consciousness (e.g., animals "perceive," but lack self-awareness). This latter idea is one of Leibnitz's useful contributions, because what he seems to be attempting to explain are the differences among vague sensations, clearer perceptions, distinct conceptions, and lucid awareness. He postulated the existence of low-level sensations or *petites perceptions* ("little perceptions"). Together they form apperception. There is, then, a continuum of consciousness. However, as later research would show, this continuum is in fact an illusion: our perceptions are cognitively but not consciously registered. What shows up on the radar screen of our interiorized "I" is not raw sensory data but cognitively processed mental images.

Despite the mysticism of his monadology, Leibnitz moved forward theories on the mind by focusing attention on the subtleties of conscious experience. We might say that Leibnitz "psychologized" the world to the degree that he foregrounded perception as central to the human condition. His *petites perceptions* foresaw the negative sensations of Fechner, the unconscious inferences of Helmholtz, the apperception of Herbart and Wundt, Freud's unconscious, and twentieth-century research on nonconscious cognition.

Immanuel Kant and the Transcendental Mind

German nativism found its supreme expression in the complex and subtle idealism of one of the world's most influential philosophers, Immanuel Kant (1724–1804). His great synthesis of empiricism and rationalism was the most ambitious attempt at the time to account for mind–body dualism generated by experiences of interiorization.

Kant reacted to interiority by pushing the mind beyond spatiality and into a transcendent realm. We might say he superinteriorized mind, positioning it even beyond perception and sensation. In this sense he recognized the limitations of empiricism and did not accept that mind was a mere bundle of perceptions. Rather, it was organized—or even better—it organized incoming sense data so that features of the objective world that appear to be "outside" us actually result from the imposition of inherent structures of thought from "inside" us. What we experience has two subjective sources. The first is sensibility ("perceptions" or "intuitions") that is passive or receptive. The second is understanding, driven by active principles of conceptualization or categories. There were twelve of these, including unity, totality, reality, cause and effect, reciprocity, existence, and necessity. Since the source of knowledge must be innate, it is the psyche itself that holds the answers.

Kant's prose is hard going and ponderous. However, his work is a systematic and methodical exposition of what had become by his time a clear recognition that interiority, possessed of principles of mental organization, had a reality all its own.[11] Kant set the groundwork for the speculative philosophical Psychology of the early nineteenth century, though his linkage to experimental Psychology of the latter part of the same century is more contested. Since he believed mind was transcendent, it does not occupy space and therefore is not measurable. Consequently, there could be no scientific study of mind. Regardless of this intellectual speed bump, during the nineteenth century physiological studies of the nervous system and experimental approaches would put the mind on the roadmap of modern life.

NOTES

1. Duchesneau, 2000, pp. 235, 220.
2. Duchesneau, 2000, pp. 218, 219.
3. Rey, 2000, p. 255.
4. Rey, 2000, pp. 255, 262, 257.
5. Rey, 2000, pp. 260, 265.
6. Wright, 2000, p. 237.
7. The "Association of Ideas" was added to the 1700 edition.
8. This analogy between chemistry and a proto-Psychology, while not completely mistaken, is a bit inaccurate for reasons that we do not have to mention here.
9. Boring, 1929, p. 181.
10. Boring, 1929, p. 198.
11. See Kant, *Critique of Pure Reason*, 1999 (1781); *Critique of Practical Reason*, 1997 (1788); and *Critique of Judgment*, 1987 (1790).

Chapter Eleven

Early Psychology

Making Visible the Contents of the Soul

THE INTROSCOPIC: THE PSYCHE AS AN OBJECT OF EARLY EXPERIMENTATION

That the mind would became a target of scientific investigation (i.e., modern research Psychology) may seem, from our own historical vantage point, inevitable. But two interrelated intellectual developments would have to first occur. One concerns "objectivity," a notion that has not always defined science. This may at first blush seem like a startling claim. But before the nineteenth century, scientists searched for "truth" (something more idealized and abstract) and "certainty" (which can be a prejudicial judgment). What was distinctive about being objective—which demands a very high level of meta-framing—was the aspiration to acquire knowledge that bears no trace of the knower. In the 1820s and 1830s, dictionaries (first in German, next French, and then English) began to use "objectivity" and "subjectivity" in the manner that we find familiar, and by around 1850 the modern sense of "objectivity" had arrived in European languages. But objectivity, as the counterpart of subjectivity, came with a price. The pioneers of science began to worry about a new kind of hindrance to the acquisition of knowledge: themselves.[1] This obstacle, a heightened self-reflexivity (due to interiorized subjectivity) was a consequence of the counterpart of increased objectivity. The subjective–objective division seemed to be a confirmation of and reinforcement of mind–body dualism.

The second development involved technological advances that radically introduced new ways of "seeing" the psychological: Researchers would press into service devices borrowed from the natural sciences in laboratories in order to measure and thereby visualize the "contents" and workings of the psyche. Incidentally, at least according to one observer, the transformation in the nature of visuality was probably more profound than the rupture that separates Medieval imagery from Renaissance perspective.[2] This novel scientific visuality illustrates how psychological experience was altered by Psychology. Introspection of the nineteenth-century variety and the instrumental measurement of psyche, then, transmuted the premodern introcosm into the modern introscope. Meanwhile, in addition to mental spatialization and introception, interiority highlighted other cognitive processes in the nineteenth century: self-narratization, self-reflexivity, self-autonomy, self-authorization, and individuation.

In this chapter I trace the emergence of an introscopic view of psyche as an object of early experimentation by looking at the contributions of nineteenth-century thinkers. But first, intellectual trends that complicated mind–body dualism are explored.

COMPLICATING MIND–BODY DUALISM

Before delineating the origins of modern research Psychology, two other intellectual catalysts behind its beginnings deserve brief mention: hypnotism and the personal equation. Hypnosis is one cousin of a larger family of scientifically unwelcome and unexplained oddities such as spirit possession, spiritualist mediums, automatic writing, glossolalia, and poetic and religious ecstasy. Interest in hypnotism during the nineteenth century reflected new questions about the mind that would drive the agenda of research Psychology.[3] Specifically, hypnotism, which for our purposes is an induced temporary suspension of conscious interiority, challenged ideas of self-autonomy, self-control, and individual responsibility. Physicians and mystics such as Van Helmont (1577–1683) and the Irishman Greatrakes (1629–83) had associated faith healing, magic, and magnetism with hypnotism. However, it was greatly popularized as "animal magnetism" (and later as "mesmerism") by the physician Friedrich Anton Mesmer (1734–1815). Influenced by Newton's discovery of gravity, Mesmer developed the notion that gravitational-like forces (animal gravitation) connected objects and living things. He used magnets to induce violent convulsions to cure entire groups of people of their ailments. During the nineteenth century, physicians such as John Elliotson (1791–1868) and James Esdaile (1808–59) had a great interest in mesmerism for therapeutic purposes. A more scientific approach to this intriguing phenomenon was pursued by the coiner of "neuro-hypnotism," James Braid (1795–1860) ("hypnotism" is an abbreviation of neuro-hypnotism) (figure 11.1). Jean Martin Charcot (1825–93), the head of the Salpêtière hospital in Paris and the "Napoleon of the neuroses," gave demonstrations before his students in which he hypnotized hysterics. One observing student was Sigmund Freud, who would eventually discard hypnotism as a therapeutic method.

The discovery of the personal equation is one of those many scientific developments whose great significance is often forgotten. This one phenomenon would link up astronomy, physiological Psychology, and nonconscious personal differences. The historical context: By observing stars crossing the graticule lines (the gridlike format formed by latitude and longitude) in transit telescopes, an individual would, through the "eye and ear" method of James Bradley (1693–1762), record stellar transit times. These measurements were important since clocks were set by them. In 1799, Nevil Maskelyne, the British Astronomer Royal, fired his assistant because their timings of stellar transits consistently diverged. The German astronomer Fredrich Bessel (1786–1846) noted that such discrepancies between measurements taken by different individuals were substantial and could not be attributed to mere mistakes. Rather, individuals were injecting a personal time lag into their measurements that had to be taken into account. The discovery of these individual differences indicated anticipation in response to a stimulus that is not inconsequential. This personal equation influenced Helmholtz's experiments (see below) and later research on reaction times. More broadly, it

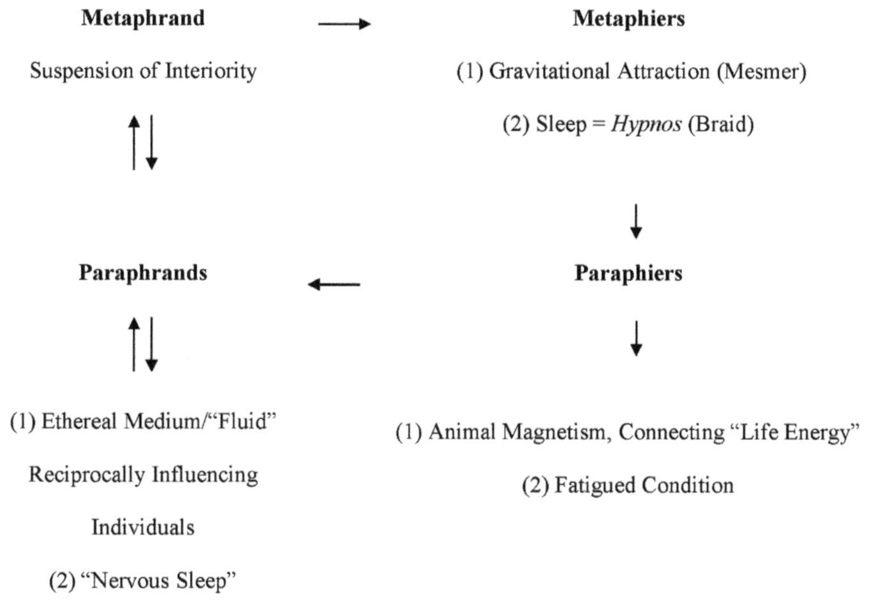

Figure 11.1. Metaphors of the Temporary Suspension of Conscious Interiority: Hypnotism.

demonstrated how thinking requires time for the processes of expectation, preparation, and attention to transpire (i.e., thinking is not mysteriously instantaneous). This all complicated mind–body dualism, and suggested a tripartite division of conscious interiority, nonconscious mind, and body (appendix D).

MODERNIZING THE SEARCH FOR THE MIND

Hermann von Helmholtz: The Physics of the Psyche

Eventually notions of the mind became less speculative and more investigatory, and the vision of human nature that was emerging was one of a psychophysical organism rather than an ensouled being. Consider the founder of the experimental study of perception, Hermann Ludwig Ferdinand von Helmholtz (1821–94). A professor of physics and physiology who taught in Königsberg, Bonn, Heidelberg, and Berlin, he believed that psychological processes were basically physiological, and physiology was a physical science. He was more in the English empiricist than the German nativist tradition, and in this sense, his work reinforced materialism. Intellectually versatile and highly inventive, Helmholtz was probably one of the greatest physicists of his day, as well as a commentator on aesthetics. His insights led to the ophthalmoscope (for examining the retina), the ophthalmometer (for examining the curvature of the optical surface of the eye), and the Young–Helmholtz trichromatic theory of color vision, upon which current theories are based. Besides vision, Helmholtz also investigated hearing. His work was put forth in *Treatise on Physiological Optics* (1856–67 [1962]), *On the Sensations of Tone* (1863 [1954]), and *Popular Lectures* (1881 [1962]).

Resonating with the nineteenth-century's focus on a clearer sense of temporality was psychophysiology's incorporation of the "fundamental dimension" of time—e.g., the time of nervous transmission, of reaction time, of attention span. "Time was the dimension that submitted mental processes to measurement; time was also the dimension that connected abstract number to concrete experience."[4] Consider how the study of neural transmission developed. In the early eighteenth century, it was accepted the nerves were so small that the senses relayed information to the brain instantaneously. The conduction rate of neural tissue was therefore impossible to measure. However, Emil du Bois-Reymond (1816–96), who worked on electrophysiology, suggested to Helmholtz that neural transmission may operate at the molecular level, making it substantially slower than the speed of light. Moreover, the invention of the chronograph (an instrument that measures time intervals) in 1849 allowed experimenters to record very short durations. By measuring the rate of transmission of the nervous system, Helmholtz realized that the sensory-to-perception process was not instantaneous but actually took time. A new technology allowed the "stretching out" of a series of temporal events that occurred between the environment and the body. An activity that once seemed mysterious and beyond human comprehension was now within the realm of scientific scrutiny.

Interestingly, the findings of Helmholtz enhanced mind–body dualism, i.e., to "separate the movement in time from the event of will that caused it was in a sense to separate the body from the mind."[5] This separation was accomplished by "unconscious inference" (*unbewusster Schluss*), a concept that many in the nineteenth century found unsettling, since it indicated that individuals routinely make decisions "out" of conscious. In other words, individuals did not appear responsible for their actions.

Physiological Psychology

Johann Friedrich Herbart (1776–1841) attempted to define Psychology, but he did so in a very circumscribed manner. An educational theorist influenced by Leibnitz, he wrote *Lehrbuch der Psychologie* (1816; English translation: *Textbook in Psychology*, 1891). His "mechanics of the mind" held that ideas obeyed Newtonian laws and interact just as physical particles do. However, Psychology was only scientific in very broad terms (as "science" was understood at the time). It could be mathematical, but like Kant, he maintained that it could never be experimental in the modern scientific sense. Indeed, for Herbart metaphysics and Psychology were the same—i.e., the fact that the qualities of interiority were difficult to grasp indicated that they were beyond complete human comprehension.

Interestingly, given Herbart's opinion that Psychology could never be experimental, he still exerted an influence and prepared the way for an empirical and experimental-based physiological Psychology. He represents, then, a "transition from the pure speculation of Kant and Fichte and Hegel to the anti-metaphysical experimentalism of Fechner and Wundt and Helmholtz."[6] What is important for our purposes is Herbart's discussion of the threshold or "limen" of consciousness. The qualities of conscious interiority were gradually coming into intellectual focus as thinkers realized what it was not, i.e., nonconscious cognition.

The brilliant Scottish anatomist and surgeon Sir Charles Bell (1774–1842) deserves credit for greatly contributing to the modern understanding of the physiology of the nervous system. The fruits of his research appeared in *The Hand, Its Mechanism and Vital Endowments as Evincing Design* (1837), *The Anatomy and Philosophy of Expression as Connected with the Fine Arts* (1872), and *The Nervous System of the Human Body* (1830). The idea that the nerves transmitted perception between the external world and the brain actually had precedent in the ancient world with Herophilus and Erasistratus (circa 250 BCE), as well as with Galen about four and a half centuries later. However, what was new was the belief that nerves do not directly transmit external impressions to the mind. Rather, what occurs in the mind results from activity in the senses. Bell experimentally demonstrated the difference between motor and sensory nerves, the muscle sense, the law of the specificity of sensory nerves, and "reciprocal innervation." He is best known for the Bell–Magendie law[7]: The ventral roots of the spinal cord contain only motor nerve fibers, while the dorsal roots contain only sensory fibers.

Johannes Peter Müller (1801–58), with his important contributions to embryology, anatomy, pathological anatomy, and the study of reflexes, is regarded by many as the founder of modern physiology (he was also one the first physiologists to utilize the microscope). However, with his investigations of perception and the nervous system, he would also lay the foundations for physiological Psychology. He discovered that the nerves of each sensory system only convey one kind of data ("specific energy or quality"). Earlier thinkers had assumed that any sensory nerve could transmit any type of sensory data to the brain. Helmholtz was one his many famous students. With Müller we are leaving a world in which sense impressions convey mental images to one in which nerve impulses carry information that are interpreted by the brain.

Even with Müller, we do not find an explicit "Psychology" autonomous from physiology or philosophy. Consider that of the eight books comprising his *Handbuch der Physiologie des Menschen* (two volumes, 1833–40) only the sixth is "Of the Mind," and this is a brief eighty-two pages. It dealt with association, memory, imagination, thought, feeling, passions, the mind–body problem, phantasms, action, temperament, and sleep. Interestingly, despite his hard-nosed scientific contributions, Müller became increasingly interested in the soul. Müller suffered several bouts of serious depression, and student disturbances associated with the Revolution of 1848 caused a serious deterioration of his health. He committed suicide in 1858.

Physiological Psychology was also developed by Rudolph Hermann Lotze (1817–81), who wrote a "medical Psychology," or "physiology of the soul" (*Medicinische Psychologie oder Physiologie der Seele*, 1852). Though he explicitly used the phrase "physiological Psychology," he was at heart a metaphysician. Concerned with space perception, he investigated how visual and tactual stimuli establish sensory experiences ("local signs") that are reassembled spatially by the mind and *within* the mind. In other words, incoming data is not inherently spatial. However, the mind is possessed with an inborn capability to arrange images spatially. This sense of spatiality is derived not from actual space, but is a function of the psyche. It is not clear from his work if he intended to highlight the *metaphoric* spatiality of interiority. In any case, Lotze's attention to how we process an essential feature of cognition foreshadowed twentieth-century advances in research on sensorimotor coordination, sensory integration, and intersensory coordination.

Gustav Fechner: The Grandfather of Psychology

The Enlightenment, in its challenge to religious authority, had chased the immortal and immaterial soul to the fringes of respectable inquiry. Nevertheless, the spiritual impulse was still strong, and brave attempts, in the face of an increasingly disenchanted and demystified world, were made to keep the divine fire lit within the individual. Sometimes these attempts mixed rationalism measured with romanticism and mysticism. By the first half of the nineteenth century, despite some important advances in physiology and anatomy, the science of the soul was for the most part still largely speculative. Indeed, the difference between the soul and the mind was still unclear, as it still is for many of us. However, psychophysics—"an exact science of the functional relations or the relations of dependency between body and mind"—was an intellectual endeavor that sought to clarify their relationship. This is where Gustav Theodor Fechner (1801–87), a philosopher, physicist, mathematician, and prophet–psychologist, enters. Fechner's psychophysics can be found is in his *Zend-Avesta, oder über die Dinge des Himmels und des Jensets* (*A Revelation of the Word,* 1851). However, his two-volume *Elemente der Psychophysik* (1860) espoused a scientific methodology that drew great public attention to psychophysics. Fechner is also regarded as the founder of "experimental aesthetics" (*Vorschule der Aesthetik*, 1876). For our purposes this development deserves passing mention because for the first time a researcher recognized how conscious interiority shapes aesthetic judgments and preferences.

In the person of Fechner science and spirituality came together in strange ways. Fechner had little time for the deterministic and cold materialism then popular among some thinkers, and his works reveal a religious, almost missionary, zeal. He blinded himself by starring at the sun through tinted glasses to study afterimages, suffered a nervous breakdown, and secluded himself in a darkened room for three years (Galileo also went blind from studying sunspots). He was painfully burdened with the overwhelming obsession that he was about to discover some great secret that he alone would have to prove using scientific methods. Eventually he recovered from his blindness and became a pantheist who confidently believed that plants possessed psyche and matter is mind.

Intent on proving that matter is imbued with mind, Fechner regarded his experiments as philosophical endeavors, not Psychological research. But ironically, in the end, his attempts bolstered a materialist Psychology. Fechner cannot be described as a psychophysical parallelist because he pursued what can be called the "identity hypothesis": mind is body (though arguably the very use of "identity" suggests an acknowledgment of dualism). Though he started with the problematic elements of mind and body, he believed their apparent dualism could be dissolved. In order to mend the cosmic fabric and reveal the underlying unity of mind and body, he believed that they must be measured and their relation determined. But how?

Fechner was inspired by the work of physiologist and anatomist Ernst Heinrich Weber (1795–1878) on "differential sensitivity" and sought to somehow equate mind with body in terms of Weber's discoveries. Weber studied how sensations were evoked by physical stimuli; more specifically, he measured the sensitivity of the skin and discovered that the relation between two stimuli is constant (a law now known as "Weber's constant"). This constant is different for each of the senses and is the smallest stimulus intensity that can be detected. The critical intensity change that is detectable is the "just-noticeable difference" (j.n.d., or the so-called "threshold of sensation"). Fechner's work would result in the "Fechner–Weber's Law" (sometimes called "Weber's Law"), that states that as sensations increase in magnitude—for example, pressure or brightness—the intensity of the stimulus increases in geometrical proportion if a j.n.d. in sensation is to occur. Additionally, if j.n.d.'s are regarded as units that are each greater than its predecessor by one, then the intensity of sensation is proportional to the logarithm of the stimulus.

Fechner also believed that "outer psychophysics" (the correlation of sensory magnitude with the intensity of physical stimulation) would eventually be replaced by "inner psychophysics" (the correlation of subjective magnitude with the central excitatory process). In other words, not just the senses but also feelings and even "higher" forms of mentation could be measured quantitatively. What Fechner seemed to be suggesting is that eventually the contents of conscious interiority could be measured in the same manner as physical stimulation.

Though many reserve the title of first experimental Psychologist for Wilhelm Wundt, Fechner may be called the originator of experimental Psychology (or more specifically, experimental quantitative Psychology). However, his advocacy of mystical panpsychism, an attempt to safeguard the spirituality not just within the individual but throughout the entire cosmos, hurt his reputation (though Fechner's belief that everything possesses features of conscious interiority is alien to many of us, we all engage in a limited, more "rational" type of panpsychism when we anthropomorphize our pets, attributing to them features of human conscious interiority). Moreover, despite the scientific advances of psychophysics during the twentieth century, it still spotlights a cosmic split and reinforces the dualistic notion that two bifurcated worlds exist, one observable, physical, and exterior, the other invisible, psychological, and interior.

NOTES

1. Datson and Galison, 2007, pp. 17, 31, 34.

2. Crary, 1994, p. 1. Crary provides a sensitive and nuanced exploration of how visuality changed in the early nineteenth century.

3. During hypnosis, one willingly abdicates one's belief about self-control and suspends belief in the metaphoric interiority believed to be encased within the head. This trancing evaporates the elements of introspectable mind–space, in particular beliefs about how "I" (active agent) controls "me" (passive recipient). The locus of agency temporarily shifts from internal–control to external–control, so that one's "I" is replaced by an outside controller (hypnotist). That beliefs about self-control are so easily arrested suggests: (1) psychic diversity; (2) psychic malleability, i.e., how easily self-authorization (decision-making) can be altered; and (3) how decision-making may be thought of as an interiorized version of social interaction between an agent and a recipient.

4. Datson and Galison, 2007, p. 264.

5. Boring, 1929, p. 43.

6. Boring, 1929, p. 250.

7. François Magendie (1783–1855), a renowned French physiologist, worked on the nervous system independently from Bell.

Part III

Time: Modern Millenarianism and Politics as "Progress"

Chapter Twelve

Meta-Framing Time

The Invention of History

We are accustomed to hoping for a future characterized by radically new sociopolitical arrangements without precedent. This is a very modern aspiration. Though notable exceptions existed, the idea that tomorrow would be better than today was not central to premodern conceptions of humankind's place in an ever-changing world. Traditionally, humankind's passage through time was cyclical or regressive rather than progressive. Hesiod, who lived in the eighth century BCE, is an example of the former view. He described in his *Works and Days* how the world of humans increasingly declined. First came the Golden Age, when people lived long and happy lives. Then came the Silver Age. Next came the war-filled Bronze Age, which was followed by the Heroic age, the time of the Trojan War. Finally, the decadent present arrived.

In this chapter I explore how the invention of history can be viewed as a meta-framing of time. This is done by introducing the different ways of defining time and a "history of history." The linkages between meta-framing of temporality and the spatializing and mechanizing of time is investigated, and how these developments led to the discovery of an objective history.

HISTORIES WITHOUT FUTURES

Rather than hoping for a better future, idealizing the past over the present has through the centuries been the common approach for dealing with change. For instance, early Judaism was "emphatically past-oriented but also antihistorical. The Scriptures were read to reinforce what the Jews already knew."

This was true for early Islam as well. Muhammad came at the end of God's plan for humankind. For Ibn Khaldun (1332–1406), the great Islamic scholar, destiny revealed itself not in temporal but rather in spatial terms. Specifically, the world was an arena of conquests by the Prophet's faithful.[1] In China, the past was privileged, and its recording was intended for constructing genealogies and justifying the Confucian cosmo-political order. The "astrologer royal" or "grand scribe" of the Han court, Sima Qian (Ssu-ma Ch'ien; 145–87? BCE), set the model for future histories of China until the fall of the Qing dynasty in 1911. Before Sima, each Chinese state used its own recording system, but Sima oriented Chinese history to the ruling house of Zhou. His successor, Pan Ku (32–92), reinforced the notion that history is an exercise in didacticism, not an impartial record of past events that have significance in themselves. During the Tang Period (618–970), "standard histories" and "veritable records" were written, but these were state property, and shared with the masses only when it suited a dynasty's purpose. Indeed, a "History Office" controlled access to the past. In India we find a complex and well-developed cyclical version of time, which had neither Alpha nor Omega. Temporal passage was repetitive and unchanging, pulsating to incessant creation, destruction, and re-creation. The basic cycle *(kalpa)* of this cosmic cycle lasted 4,320 million earthly years. Each *kalpa* contained fourteen *manvantara*s (306,720,000 years each; each *manvantara* contained seventy-one epochs or *mahayuga*s (a thousand make up one *kalpa*; and each *mahayuga*s had four *yugas* (4,800, 3,600, 2,400, and 1,200 years, respectively).

THE HISTORY OF HISTORY

How do we describe the history of history? In the Western tradition, we look back to personages such as Hecataeus of Miletus (circa 540–489 BCE), who wrote the *Genealogiai* (or *Historiai*), a detailed account of the Greeks and their beliefs. He is known to us today because he attempted to distinguish myth from historical fact. Then there is Euhemerus (circa 300 BCE), author of *Sacred History*, who theorized that the gods were originally distinguished historical personages (thus euhemerism). The Athenian Thucydides (circa 460–circa 400 BCE), famous for his *History of the Peloponnesian War,* attempted, for the standards of the day, an "objective" history. Herodotus of Halicarnassus (circa 480–425 BCE), the "father of history," wrote in a realistic, rather than a poetical, mode, as seen in his *History*, which is an account of the Persian Wars.

Lacking a modern sense of the historical, these ancient works, despite their impressiveness, were not history as we consider it. Of course, if loosely defined, something like "history" did exist in the ancient world. However, this earlier conception of the passage of time centered on the recording of royal deeds, tracking sacred calendars, and retelling cosmic events. Certainly a sense of pastness can be found in early works, but often it was the *universal* that was significant, expressed in emotionally moving mythologized accounts rather than common prose (this was a time, after all, when the function of poetry was to store what was memorable to the community). The *particular,* such as commonplace facts, ordinary events, or the mundane customs of foreigners, was less important. History was not linear, but contemporaneous; no notion that contemporary societies were fundamentally different from those of the past is evident. It would take many centuries for the idea to develop that history is a strictly linear process that needs to be objectively recorded for its own sake.

St. Augustine (Aurelius Augustinus, 354–430; Bishop of Hippo from 395) was an interesting exception to the lack of historical linearity. Writing during the decay of the classical Western world, he is noteworthy because he dismissed the pagan idea of historical cycles (*circuitus temporum*). In *The City of God* he offered a spiritualized chronology with the birth of Christ as a singular event that bifurcated human history into a "before" and an "after." Nevertheless, his was not a modern view of "progress" driven by technological innovation with rapid political adjustments to economic improvements. Note that the six "world periods" (*aetates*) used by the early Christian thinkers to divide time before the coming of Christ were not historical but "prophetic and theological" or "stages toward the Incarnation."[2]

It is important, then, to note that "neither the Islamic, the Black African, the Hindu nor the East Asian Confucian world seems to have discovered the future as a new place, attainable but never visited before."[3] Nevertheless, the idea that a better future awaited us would inform Christian and Islam theology. Indeed, while the great religious traditions of Hinduism and Buddhism "sought ways *out of* history, Christianity [and later] Islam sought ways *into* history."[4] The idea, eventually desacralized, that a better future awaited us would eventually shape our modern faith in progress.

SPATIALIZING AND MECHANIZING TIME

For our premodern predecessors, time, like other aspects of reality, had "humanly comprehensible dimensions," and though time was impressively expansive, it did not exceed the soul's ability to conceive of it. And "infinity was the antithesis of time, rather than its extension." In this older view, temporality carried great religious and symbolic value, but it had "little to do with precision."[5] However, in addition to viewing it as linear, the modern mind characterizes time as "empty," an abstracted, meta-framed spatiality waiting to be filled in by the rush of events coming from the future rather than a place already occupied by past occurrences.

The regular measurement of time is not cognitively inherent to human experience, though its use for calculation in economics (whether agricultural, industrial, or postindustrial) and to provide religio-cosmological order is universal. Around the globe all civilizations have treated time as a key piece of the cosmic fabric. Their conceptualizations were actually quite sophisticated, and their understandings of sacred astronomy and calendrics became crystallized in massive stone works that, silenced by time, still stare at the heavens. The meaning of the future is apparent in the crucial difference between schedules and calendars: "Scheduling cultures do not commemorate; they plan. They are not interested in resurrecting the past but in manipulating the future." In this "new time frame, the past is merely a prologue to the future."[6] Calendars merely chronicle.

Through the metaphoric spatializing of temporality within the mind, time "took on the character of an enclosed space: it could be divided, it could be filled up, it could even be expanded by the invention of labor-saving instruments."[7] The ontogenetic development of the individual's inner sense of time or a "narrative self" corresponds to the historical emergence of temporal linearity at the phylogenetic or ideo-cultural level: at first unproblematic presentness exists. Next, a notion of the past develops, and then eventually a more explicit, concrete, and complex view of pastness comes into focus. Finally, the notion of the future emerges. Abstract time is also implicated in new forms of cognition that were emerging. A new vacuous "space" was invented.

Ever since the advent of writing time has been spatialized and meta-framed via technology: "Record events in chronological order on parchment or paper and you have a time machine. You can step back and forth and observe beginnings and endings simultaneously. You can alter time's direction, and you can halt time so as to examine individual events."[8] Such spatio-technologizations of temporality have become more advanced over the centuries, but with every innovation, the representation of time by conscious interiority has become more abstract, spatialized, and meta-framed.

THE CLOCK AND THE SPATIALIZATION OF TIME

In the Western world the regular measurement of time originated within the walls of the monasteries where prayer demanded the regulation and regularization of time. In such surroundings "surprise and doubt and caprice and irregularity were put at bay." The instrument that would be invented to regulate and measure events, of course, was the clock. Eventually, mechanical versions of the clock would appear in the Western world during the thirteenth century. By the 1300s, clock towers had become common in European towns, and belfries ushered in a mentality configured by tolling bells sounding the hour. The clock made time "external, autonomous, continuous, exacting, quantitative, and divisible."[9] Eventually clocks became portable and could be placed in households. Driven by the disciplining demands of commercial society that decreed that time is a value in itself that can be obtained or lost, clocks took a more personal form as pocket watches and wristwatches.

Time, increasingly secularized, became something shared and public. The story of the clock's rise in the West is the "story of new modes and widening arenas of publicity." This understanding of time did not have to follow this trajectory. Consider China, where impressive mechanical clockworks of great sophistication were developed. Time itself was almost a type of property owned by the authorities, and dynasties were symbolized by their official calendars that linked them to the cosmos. "Private calendar–making was punished as a kind of counterfeiting—as both as threat to the security of the emperor and as an act of lèse majesté."[10]

The clock also dissociated time from happenings in the human world and thereby created the belief in an "independent world of mathematically measurable sequences," i.e., science.[11] But economics should not be neglected as a motivating factor in wedding time to mechanics and something measurable. The calculations of capitalism—profit as hypothetical gain, bills of exchange, speculation, etc.—changed the notion of time by encouraging individuals to imagine a different future filled with wealth. The techniques of science were not established until after Galileo and Newton. However, finance arose in its current form before the age of the machine. The merchant and banker Jakob Fugger (1459–1525) and the financier, banker, and philanthropist J. P. Morgan (1837–1913) could "understand each other's methods and point of view and temperament far better than Paracelsus and Einstein."[12]

In *Technics and Civilization* (1934), Lewis Mumford highlights the clock's impact, writing that it, and *not* the steam-engine, was the defining machine of the modern industrial age. More than coal, iron, or steam, the clock was indispensable to industrialism. Specifically, in its "relationship to determinable quantities of energy, to standardization, to automatic action, and finally to its own special product, accurate timing, the clock has been the foremost machine in modern technics." It "marks a perfection to which other machines aspire."[13] Indeed, the universe itself would be envisioned as a great cosmic clock, and the universe-as-watch metaphor would become influential in the early modern period, though the analogy of the universe as a giant clockwork wound up by God was introduced by Bishop Nicole d'Oresme (1330?–1382).

DISCOVERING OBJECTIVE HISTORY

"History" was a fuzzy mix of legend and myth, or perhaps paeans to patrons. It was used by the religio-political elite to justify the sociopolitical order. Or it was meant to be studied in order to confirm faith in God's plan for humankind. To critically investigate the past or make claims to objective authenticity invited, if one were lucky, censure, and if one were not so fortunate, torture or burning at the stake. The emergence of a modern sense of history was a monumental development. It is difficult to point to a series of dates when history came into existence, but certainly, by the late Medieval or the early modern period it was well on its way. Individuals of the Renaissance would indulge their "fantasy of reviving the classic past of reliving the splendors of the Roman civilization: the cult of history, appearing first in daily ritual, finally abstracted itself as a special discipline."[14]

The metaphoric spatializing of temporality into grand eras that has a beginning and an end was underway by the time of Flavio Biondo of Forli (1392–1463). He spatialized the thousand years between the fall of Rome and his own time into a single "middle" era, suggesting the tripartite division of "ancient glory," "modern rebirth," and a middle age of "disintegration and decline in between." Though Biondo himself never used the Latin *medium aevum* ("Middle Ages" or *media aetas*), later historians would employ his ancient, Medieval, and modern division.[15] By the last third of the fifteenth century, a distinction between "ancient" (past) and "modern" (now) was being clearly made.

Obviously, a practical political motivation behind the sharp delineation of temporal sequencing existed. Lineages, inheritances, and claims of jurisdiction were often justified with forged documents, but with a narrativity of events or evidence of objective chronology (or their lack) on the side of a claimant, forgeries could be challenged (a diploma—from "doubled," or "folded"—described documents folded not only for storage but to keep their contents secret since the political authorities possessed property rights over their contents). For instance, a pioneer in searching for objective truth in the past was Lorenzo Valla (1407–1457). In *Treatise on the Donation of Constantine* (1440) he proved that the Donation of Constantine was a forgery. This had major political repercussions, since for centuries the Donation had been used by popes to claim dominion over Rome and the whole of Western Europe. In *Annotations on the New Testament* (not printed until 1505) Valla also applied critical history to the Bible.[16]

What modern individuals take for granted, "vistas of contemporaneity" or "a sense of what was going on all over the world at the same time," began to appear in the early modern period. Consider the French scholar Joseph Justus Scaliger (1540–1609), who in "Correct System of Chronology" brought together all the major events of antiquity into a single time line. Impressively, he relied on his knowledge of Hebrew, Chaldean, Arabic, Phoenician, Ethiopic, Persian, Syrian, as well as philology, mathematics, astronomy, and numismatics. Eventually idioms were developed that partitioned time into "ages, "eras," and "epochs." These currently used divisions—antiquity, the Middle Ages, Renaissance, the Enlightenment, modern, postmodern, etc.—united humankind in a grand community of historical continuity that transcended city, region, and nation.[17]

NOTES

1. Boorstin, 1985, pp. 566, 567, 560.
2. Boorstin, 1985, p. 611.
3. Therborn, 1995, p. 19.
4. Boorstin, 1985, p. 566, emphasis in original.
5. Crosby, 1997, pp. 22–23, 23, 35.
6. Rifkin, 2004, p. 106.
7. Mumford, 1934, p. 17.
8. Crosby, 1997, p. 227.
9. Rifkin, 2004, p. 110.
10. Boorstin, 1985, pp. 73–74.
11. Mumford, 1934, p. 15.
12. Mumford, 1934, p. 23.
13. Mumford, 1934, pp. 13, 17, 14, 15.
14. Mumford, 1934, p. 17.
15. Boorstin, 1985, p. 580.
16. Boorstin, 1985, p. 576.
17. Boorstin, 1985, pp. 596, 600, 611.

Chapter Thirteen

Liberating the Psyche

The Emerging Faith in Progress

THE MARCH OF SECULAR–HUMANISTIC PROGRESSIVISM

Regardless of the important differences among today's different societies, they arguably all share, at least to some degree, a faith in the same religion. This religion lacks a formal name and defies easy characterization. However, we can at least delineate its basic dogma. To ignore the significance of this faith is tantamount to overlooking the most prevalent, successful, and deeply rooted belief system history has ever seen. Whether one is a Christian in North America, a Jew in Israel, a Hindu in India, a Muslim in Indonesia, or a Buddhist in Thailand, one is a believer. Even atheists are among the faithful.

In this chapter I will discuss and give a name to the faith that characterizes so much of modernity: secular humanistic progressivism. Progressivism is secular because it has been suspicious, if not downright hostile, to religion. However, not all of secular humanistic progressivism necessarily objects to religion. Indeed, some of its variants see an important role for spirituality in modern life, and the separation of church and state is but one way in which a mutual accommodation has been reached. Progressivism is humanistic because it is "human-centered" in the sense that the purpose of life has become the gratification of the expanding desires agitating within the individual's interiorized psyche.

An understanding of any society requires more than just a passing glance at its explicit, formalized politico–economics, cultural norms, and religious thought. These forms of organized knowledge are more or less obvious. A more fundamental appreciation demands an investigation of a society's mythical foundations, buried deep beneath its everyday conceptualizations and grander visions; what is needed is an exploration of its "invisible institutions," its implicit forms of understanding that everyone somehow knows but rarely questions in any sustained, meaningful way. In any case, many reasons explain the unquestionable nature of secular humanistic progressivism. Having obtained mythic proportions so deep, few probably feel the need to challenge its orthodoxy. However, a more convincing reason is that secular humanistic progressivism provides a utopian promise, and to that degree, it is imbued with spirituality, no matter how rationalized and scientized that spirituality may be.

Of course, as members of any religion might do, people occasionally question some of its articles of faith. However, very few systematically challenge the entire creed. Specifically, many studies in political science and economy suffer from the view that today's politico–economic ideologies and the national structures they legitimate are not mythical, but rather "rationalized," "efficient," and "modernized" arrangements. The flaw with this view is that rationality, efficiency, and modernity are historically specific, culturally informed, and socially constructed; they are normative judgments. These are not aspects of some universal truth, nor are the results of inevitable evolution from simpler to more complex forms. The religion in question, which is of a natural rather than a supernatural variety, was a long time in the making. There was nothing inevitable about the emergence of this modern faith; its appearance was dependent on many intellectual currents, and it is composed of different tenets.[1] Some historical perspective on the faith in "progress" should demonstrate its contingent nature.

"PROGRESS" AND SOCIAL RE-ENGINEERING

It is difficult for us to imagine societies that lacked a clear notion of the future—in the sense of anticipating a different, improved society. However, before the early nineteenth century, time was not clearly linearized. Once a conception of the "future" became accepted by the late eighteenth century, it became possible to envision improvement of society and "progress." If a somewhat vague faith in progress existed in the seventeenth and eighteenth centuries, this conviction becomes more specific and explicit by the nineteenth, often bolstered by the science of the day or driven by political and economic aspirations. The notion of time had undergone a profound alteration. Indeed, one of the most important changes that ushered in modernity was the redefinition of time. This new understanding relied on a meta-framing of history itself. This notion is intimately related to the destabilization of socioeconomic structures and, since the French Revolution of 1789, revolutionary sentiments. The political economics of stasis and stability gave way to one of transformation and constant renovation of social institutions. The demands of industrialization, constitutionalism, national state centralization, and empire-building were all mobilized in the name of progress.

SECULARIZING AND DESACRALIZING TIME

For centuries time was stuck in St. Augustine's theological "perpetually present": In "*eternity* nothing passes by; everything is present, whereas *time* cannot be present all at once."[2] Though people wrote "histories," these were of limited, local significance, and time was marked by the regnal year of a ruler or dynasty. "AD" (*anno domini,* "the year of our Lord") was devised in 525 by the astronomer–mathematician monk Dionysus Exiguus, and over the next several centuries was gradually adopted ("BC" would not be used until much later in the seventeenth century when a more solid sense of the past developed). Eventually, time-keeping passed into time-serving and time-accounting and time-rationing, and "Eternity ceased gradually to serve as the measure and focus of human actions."[3] The past had been (re)discovered in Europe, especially as represented in classical works. An interest in "practical knowledge"—e.g., mathematics, medicine, law—played a role, but also "a search for new attitudes, ideals, and elegances" became more pronounced. By the twelfth century the notion that the key to understanding the world lay "not in the natural order, but in the history of man" had had an impact. Nevertheless, although the twelfth century was "passionately interested in history," a critical historical sense would not begin to appear until the 1700s.[4]

EARLY NOTIONS OF PROGRESS

The idea that society is progressing can be traced back as far as the ancient Greeks. However, the idea of progress as something *central* to politico–economic theorizing, as well as evidence of how it has saliently impacted the average person's worldview, are very different issues. In any case, examples of pre-Enlightenment thinkers who argued for what would later be called a "social evolutionist" view do exist. For example, the Valencian scholar and humanist Juan Luis Vives (1492–1540) held an evolutionary theory, believing that language was necessary for the emergence of society which arose through cooperation from the need of subsistence. Jean Bodin (1530–96), important for his efforts in studying politics objectively, also believed in a type of social evolution. He postulated that human history passed through three periods: (1) religion; (2) warfare; and (3) inventive skill. Bodin also noted differences in cultures and examined the influence of the natural environment on social and political systems.

At about the same time, the publication of Sir Francis Bacon's (1561–1626) *The New Atlantis* (1627) had a profound impact on eighteenth-century thinking about progress. This book, significant because it viewed knowledge in a radically new light, described a community on the imaginary island of Bensalem and its research institute called "Solomon's House." For Bacon, the purpose of accumulated knowledge was to increase human happiness. This is a decidedly modern idea, steeped in an interiorized view of the human condition that privileges emotional well-being, rather than a religious orientation that saw the primary purpose of life as spiritual salvation. We are moving closer to another major historical watershed: the establishment of government (or property management regimes) amenable to constant, future-oriented change so as to satisfy the interior life of the individual—"life, liberty, and the pursuit of happiness." Thanks to the advances of the Industrial Revolution, Bacon's writings would go from the imaginary to the attainable, and as we enter the Enlightenment, arguments for social progress become more explicit. Bacon's thinking influenced Denis Diderot (1713–84) and Jean Le Rond d'Alembert (1717–83), who credit Bacon for inspiring their *Encyclopedie* (1751), a mammoth dictionary of the arts, sciences, trade, and manufacturing.

By the eighteenth century, the directionality of time had become a category of thought in its own right. Sir Isaac Newton (1642–1727), besides the scientific and mathematical discoveries for which he is famous, accepted 4004 BC as the date of creation and wrote extensively on the relation between theology and temporality. Thinkers such as Bernard le Bovier de Fontenelle (1657–1757) made the idea of progress even more explicit in his *Dialogues of the Dead* (1683). Fontenelle was a philosopher of mathematics and science, and in *Dialogues* he described an imaginary dialogue between Socrates and Montaigne in which the latter is told by Socrates that history naturally progresses. Fontenelle's *Digression on the Ancients and Moderns* (1688) articulated an even more explicit idea of progress that concerned the "*querelle des anciens et des modernes*" ("the debate of the Ancients and Moderns"). The latter had rebelled against the Renaissance (which had idealized the Greco–Roman classics) and positively accepted advancements in science. Fontenelle argued in favor of progress with respect to the arts and the sciences, arguing that each age has the advantage of not having to rediscover what was accomplished in previous times. "Progress since the past" and "progress for the present" were conceivable, but Fontenelle extended such thinking to "progress into the future," or the unseen realm of the not yet.

In *The New Science* (1725), Giambattista Vico (1668–1744) wrote about the social—in addition to the supernatural—forces behind historical change. He believed in the principle of *corsi e recorsi* ("course and recourse") or the "the law of historical recurrence" that was guided by God. For Vico, three stages characterize humankind's evolution: (1) the age of the gods; (2) the age of heroes; and (3) the age of men (these ages were commonly mentioned in the ancient world, as in the work of Herodotus; circa 484–420 BCE). Vico was also influenced by the ancient Greek idea that history moves cyclically from monarchy, to theocracy, aristocracy, democracy, and back again to monarchy. Though cyclical, history followed an upward spiral. However, regardless of the role of Providence, Vico, whose thinking was at once mystical and modern, also believed that people themselves establish and make their own societies. This was a revolutionary idea and highly significant for the future development of the social sciences. What is important about his work is the explicit description of the forces of change; in other words, social institutions were human creations and modes of thinking that followed a trajectory mirroring humankind's changing experiences.

After the pastness of classical antiquity was explicitly recognized, increased spatialization of temporality within the imagination allowed the pushing of the past even further back into time. The great naturalist Georges Louis Leclerc Comte de Buffon (1707–88) in his *Les Epoques de la Nature* (1788) suggested that the earth was far older than the biblical date of creation, i.e., 4004 BC. Charles Darwin (1809–82), of course, would later also challenge this religious orthodoxy. Pushing time further back led to the discovery of pre-history. The Danish businessman and numismatist, Christian Jürgensen Thomsen (1788–1865), authored *Guide to Scandinavia Antiquities* (1836) in which he outlined a prehistory partitioned into a three-age system: Stone, Bronze, and Iron. In 1819 he opened a museum in which artifacts were arranged according to what he called the "principle of progressive culture." A student of Thomsen, Jens Jacob Worsaae (1821–85), would develop these ideas and earn the title of the "first modern archaeologist."[5]

Anne Robert Jacques Turgot (1727–81), France's minister of finance (1774–76) and physiocratic economic reformer, also argued that societies are improving through time. In *Plan de Deux Discours sur I'Histoire Universelle* (1750) he examined how factors familiar to modern social science, such as diffusion, migration, population density, and economic productivity, led to progress. Turgot believed that culture is learned, is cumulative, constitutes a whole, is transmitted from generation to generation by means of language, and that new knowledge is built upon things learned in the past. Before the idea of progress, knowledge was something to be rediscovered and recovered from the past, not discovered.[6] Thinkers such as Turgot would change this perception, and for this reason he in no small way contributed to the basics of modern thought.

Around the same time, Marie Jean de Condorcet (1743–94), a friend of Turgot's, wrote a paper called "Fragment on Atlantis." Inspired by Bacon's idea for a research institute (Solomon's House), Condorcet suggested the establishment of a world research center. A brilliant mathematician, Condorcet asserted that history had progressed through ten stages in his 1795 *Sketch for a Historical Picture of the Progress of the Human Mind*: (1) hunting and fishing; (2) pastoralism; (3) agriculture; (4) commerce and science in Greece; (5) Roman Empire; (6) from the fall of the Roman Empire to the Crusades; (7) from the Crusades to the invention of printing; (8) from the invention of printing to the Enlightenment; (9) from Descartes to the foundation of the French Republic; and (10) from the French Revolution onward. Unlike Turgot, Condorcet had a more mystical concept of progress, and viewed it as an independent power or force, not necessarily related to the earthly phenomena of diffusion, migration, population density, or economic productivity. Nevertheless, Condorcet has been highly rated as a forecaster of the future. The last of the *philosophes,* he was hounded by the terror of the French Revolutionaries, arrested, and died in prison. We might also mention Madame de Staël (1766–1817) who, influenced by Condorcet, wrote *On Literature Considered in Its Relationship to Social Institutions* in 1800. This contained a theory of progress and in a sense "discovered" society in its attempt to contextualize art within changing society.

The thinking of Wilhelm Friedrich Hegel (1770–1831) has left an indelible mark on nineteenth-century philosophy. Indeed, Hegel is considered by many to be the most influential Western philosopher. Perhaps more than any other modern thinker, the idea of progress played a pivotal role in his ruminations. In his many writings, particularly his *Lectures on the Philosophy of History* (1837), the various topics he treated revolve around notions of unfolding, growth, and progress. For Hegel, the world was imbued with a certain directionality, and the dynamic power of history was its own spirit, a restless dialectic animated by a mystical "reason" above but guiding human society.

Of all the thinkers in the Western tradition it is Hegel who seems to have intuited that Psychology, politics, and history are cut from the same cloth. Though I will not pursue it here, it appears that Hegel (1770–1831), though relying on an idiom that is at times impenetrable, recognized that the psyche and its changes cannot be understood outside historical context. Indeed, my thinking on interiorization and Hegel's view that world history is the progress of the consciousness of "freedom" resonate in important ways (though such a claim must be made cautiously, given the alleged profundities embedded in his abstruse German).[7] Hegel admitted that "freedom" was an "incalculably vague term," but for him it was an essential aspect of how "mind" or "spirit" (in German *Geist* means both) works its way through history and appears to itself (self-reflexivity). We can, of course, discard the teleological and mystical aspects of Hegel's arguments and look for techno–economic developments that account for the general historical patterns that he noted.

THE EMANCIPATORY ETHOS

The Groundwork for Modernity

Despite all the methodological problems surrounding "modernity," can we nevertheless find a common conceptual denominator to the dominant narratives? I submit we can: A belief in the inherent benefits of unshackling people from social strictures, thereby releasing human potential rooted in the hidden realms of the interiority of the individual. Such unshackling took various forms depending on local ideo-institutional circumstances and affected different groups—caste, occupations, status groups, religious classes—in distinctive ways. However, this unleashing can be summarized by the term "emancipatory ethos" which is comprised of a list of major developments associated with modernity:

Capitalism sanctioned the accumulation of private wealth and the commodification of new forms of value.

Science set humankind free from the limitations of the natural world and developed liberating knowledge forms.

Technology allowed the emergence of new inventions geared to economic growth.

Industrialism released innovative technologies that satisfied the demands of a growing mass society.

National Statism liberated entire peoples from imperial and foreign domination.

Market Forces let loose and democratized monetary exchange systems unchained to the avarice of officialdom.[8]

Democracy allowed popular participation to accompany bourgeois economic clout.

Bourgeois Lifestyle permitted regimes of property accumulation that led to a new type of wealth-based status aspiration.

Civil Society granted an impartial space free from both officialdom and the obligations of kinship and the private realm.

Consider, as an example of the emancipatory ethos, the worldwide trend of the increasing number of "liberal democracies": 1790 = 3; 1848 = 5; 1900 = 13; 1919 = 25; 1940 = 13, 1960 = 36; 1975 = 30; 1990 = 61. There have been, to be sure, serious setbacks and reversals, but the general trend is undeniable. Why? Exploring the particulars of the political situation in a certain locale surely provides some answers, but I suggest that global trends which transcend place, such as increasing self-autonomy and other features of interiorization, played an essential role.

With all this stated, the premodern/modern divide can be characterized as a transition from a sociopolitical environment that discouraged mobility to one that encouraged it, from the assumption that individuals require restrictions on their behavior to the belief that freedom itself is a natural good. Suspicions toward religion (generated by the horrific Thirty-Years War, 1616–48) and toward monarchy (epitomized by the revolutions in England during the seventeenth century) laid the intellectual groundwork for widespread skepticism. The late eighteenth and early nineteenth centuries witnessed the impact of the Enlightenment (a decidedly secular attitude toward nature and the "discovery of society"), as well as two massive movements of emancipation that eventually swept across the Western world, one political the other economic, of which the French Revolution and the Industrial Revolution are emblematic. By the mid-nineteenth century "liberalism" had become the catchword for emerging types of freedoms brought on by economic dislocations and new classes generated by the Industrial Revolution. However, the pursuit of liberties took on forms that often clashed: "Economic" liberalism (laissez-faire marketism that spawned exploitative working conditions) did not sit well with "political" liberalism (individual rights, constitutionalism, antimonarchical republicanism).

As an example of liberalizing forces outside Europe, consider the politico–religious utopianism of the Taiping Rebellion (1850–64), a bloody peasant uprising led by the quasi-Christian and mystic Hong Xiuquan (1814–64). Originating in southeastern China, the Heavenly Kingdom of Great Peace was a state within a state that dominated the Yangtze River Valley and almost toppled the Qing Dynasty. Though still very much beholden to premodern Confucian moralism, the Taiping leaders presaged twentieth-century revolutionary developments in China and inspired later revolutionaries, such as Sun Yat-sen and Mao Zedong. During the Taiping Rebellion an estimated 20 million people died, and huge swathes of the fertile Yangtze River valley were laid to waste. Pertinent to our own topic is how this upheaval resonates with elements of modernity and the emancipatory ideologies sweeping nineteenth-century Europe: proto-nationalism, economic egalitarianism, new forms of social organization and mobilization, new conceptions of the person, as well as other progressive ideas, such as sexual equality.

NOTES

1. For a discussion of the sociology of time, see Zerubavel, especially his discussion of "Private Time and Public Time," 1981, pp. 138–66.
2. Augustine in Warner, 1963, p. 265, emphasis added.
3. Mumford, 1934, p. 14.
4. Morris, 1972, pp. 11, 52–53, 54.
5. Boorstin, 1985, pp. 606, 607–10.
6. Crosby, 1997, p. 61.
7. Given the resonances between my thinking and Hegel's, I had originally planned an entire chapter on Hegel's *The Phenomenology of Spirit* (1803) and *Lectures on the Philosophy of History* (but that will have to wait for a follow-up project).

8. Contrary to much commentary on the subject, states did not withdraw their control over economics. Rather, they redesigned and redirected the exchange circuitry so as to better accommodate the commercial classes, the rising bourgeoisie, and those who came to be called the capitalists. After all, without infrastructures built, maintained, and protected by states, the exchange circuitry of the market simply cannot circulate wealth.

Chapter Fourteen

The History of Humankind

Climbing the Ladder of Civilization

In the previous chapter we looked at belief in inevitable progress, a faith rooted in the notion of unilinear time. This belief, that societies are moving toward a brighter future, is a relatively recent mind-set that occurred later than many of us might assume.

This chapter continues themes from the previous chapter by examining how a meta-framed history led to the secularizing and desacralizing of time. This development allowed the emergence of a belief in "progress" and its concomitant ideas of social re-engineering. This argument is made more specific by exploring the consequences of an emancipatory ethos that laid the groundwork for modernity and other nineteenth-century intellectual developments.

TOWARD A BRIGHTER FUTURE

For all their intellectual innovation, thinkers of the Renaissance and Reformation still had their eyes on the past. Even the Enlightenment, some aspects of the French Revolution, and the elite of the British industrial Empire also looked backward to ancient models. Modernity "looks at the future, hopes for it, plans for it, constructs it, builds it." It is a temporal perspective "turned to the future, conceived as likely to be different from and possibly better than the present and the past." Modernity, then, is really about discovering the future.[1] The idea that time moves like an arrow—with a past, present, and future—is so much a part of our worldview that most of us believe it to be an aspect of the natural world. However, time, regardless of its use as a variable in physics, is historically specific and culturally constructed. Indeed, time has been conceived in various ways in different periods and places, with cyclical concepts being prevalent. The Judeo–Christian eschatological tenet that humankind is living through a temporal reality—God's plan for the world—that has a distinct beginning and end, has become one of the most powerful intellectual orientations of modernity. In particular, Western scholarship since early modernity has centered its worldview within the unilinear temporal paradigm, and this proclivity culminated in the eighteenth and nineteenth-centuries' obsession with the search for origins, stages, and phases of development. What is important for our purposes is how the notion of progress, whatever else it may be, can be understood as the belief that entire societies can meta-frame their way up through the stairs of the ages.

PROPHETS AND PRIESTS OF PROGRESS

When we reach the historical period of Saint-Simon and Comte, we encounter thinkers who had moved from abstract speculations to more action-oriented (though in retrospect, naïve) proposals and programs advocating social progress. Saint-Simon and Comte, who had worked together for seven years, must be seen in the context of the Industrial Revolution and the spread of Enlightenment notions of an emancipatory ethos. In their writings, both blended a strange religiously inspired conservatism with a strikingly modern appreciation of the economic implications of techno–scientific developments. And both were original thinkers whose ideas would have a lasting impact on the future development of the meta-framed corpus of knowledge constituting the social sciences.

SAINT-SIMON: PROPHET OF INDUSTRIALISM

Count Henri de Rouvroy de Saint-Simon (1760–1825), as his name suggests, was born an aristocrat. A colorful character, he advocated democracy (though not necessarily equality and what we would call individualism) and fought in the American Revolutionary War. He believed that progress is based on science, and that through new inventions society could be improved. For Saint-Simon, since science was a means to improve society, it was therefore a moral and sacred endeavor, and he viewed what was then called "industrialism" as a new order whose emergence would radically alter human relations. Indeed, his hopes and visions for humankind are prophetically similar to what modern society has become. In his voluminous writings, he appreciated the role of the common workers, or *les industries*, and stressed the necessity of practical work while taking swipes at idle aristocrats. For Saint-Simon, society was a factory, and government should devote itself to efficiently managing this workshop writ large, not merely ruling over it. Though Saint-Simon envisioned the ideal society as a harmonious hierarchy, he also believed it should be meritocratic.

Saint-Simon's *New Christianity* (1825) reflects the utopian character of his thinking. He had plans for an international council that would rule Europe (despite the obvious differences, one is tempted to compare this council to the EU). This parliament-type body would be composed of three houses, one for: (1) inventions, composed of scientists, inventors, architects, painters, poets, writers, and sculptors; (2) examination, composed of physicists, mathematicians, and other scientists who would critically evaluate the first house's plans; and (3) bankers, industrialists, and businessmen who would execute the projects of the first house. In order to gain knowledge for running the world, Saint-Simon saw the need for a scientific study of society. He described this intellectual endeavor in a metaphoric expression that signaled his view of society as an interconnected living entity—"social physiology."

During his lifetime, Saint-Simon and his follows proposed ambitious plans for canals, railroads, and steamship lines. After his death, his disciples organized a semireligious Saint-Simonian movement with branches in various countries that propagated his ideas on social reform and progress. Indeed, after his death, many of his follows became famous financiers and industrialists, and one, Ferdinand de Lesseps (1805–94), even realized one of Saint-Simon's dreams—the Suez Canal.

Many of Saint-Simon's ideas about the nature of society would be adopted by Comte and later thinkers in the social sciences. "The future belongs to us," Saint-Simon was said to have told his followers on his deathbed. In many ways he was right. Our world is now run by elitist bureaucrats, technocrats, and economic experts. Their goals are all grounded in an unquestioned faith in the need to somehow improve the human condition, as is the pursuit of happiness by the toiling and laboring masses. Though his visions about industrialism do not seem radical to us now (indeed, they are accepted as commonsensical), it must be remembered that few even attempted to imagine how the massive economic changes taking place at the time would fundamentally restructure the social order.

COMTE: SOCIETY AS A SCIENTIFIC OBJECT

Auguste Comte (1798–1857) is usually credited with founding "positivism," the epistemological stance that the world should be studied by only looking at that which is observable and verifiable. However, this perspective would soon be taken up by those who believed that science has all the answers for improving society. Comte, who was Saint-Simon's secretary until they had a falling out, was well trained in the sciences of his day. Like Saint-Simon, he was a prescient observer of the trends and direction of his times, and shared with all great thinkers of his time a millenarian vision. He believed that humankind had passed through three stages of knowledge: (1) theological (explanation by spirits, gods); (2) metaphysical (explanation by abstract philosophical speculation); and (3) positive scientific explanations based on observation, experiment, and comparison. He expressed his ideas in his massive *Course in Positive Philosophy* (1835–52) and *System of Positive Polity: A Treatise of Sociology, Instituting the Religion of Humanity* (1851–54). Comte, like Saint-Simon, advocated industrialism and reserved a special place for scientist–priests in his vision of the utopian society.

But if Saint-Simon was a hard-nosed atheist and materialist, Comte was more conservative and interested in the role of faith and feeling in holding society together. He was, after all, a great admirer of the Middle Ages and the Medieval church (he was born into a Catholic family though he himself left the church). Indeed, Comte's positivism has been called Medieval "Catholicism minus Christianity." Eventually, this Christianity without the superstitions became a cult, a "Religion of Humanity" which venerated society itself as the "Great Being" and developed a calendar with days of devotion for scientists, poets, saints, and philosophers. Leaders of the movement met in a council under Comte, the High Priest of Humanity. Comte's faith attracted adherents from the United States, Russia, and other countries. Through the help of some of his follows, his motto, "Order and Progress" (*Ordem e Progresso*), is visible on the Brazilian flag. Like Saint-Simon, Comte viewed society objectively and abstractly. He termed what he considered the new science needed to bring order to the world "social physics," a metaphoric turn of phrase inspired by the physical sciences. Later he coined the word "sociology," which he divided into "social statics" and "social dynamics," or the study of social order and social change, respectively.

Comte, as well as Saint-Simon, significantly influenced the "father of sociology," Émile Durkheim (1858–1917) and French sociology (which would shape early British social anthropology) with his ideas of society as a "fact" existing ("*fait social*") outside and above the individual. His other influential ideas included: The metaphor of society as a biological organism; how society possesses "structural" components that "function" in concert; and a consensual view of how human sentiments hold society together, i.e., "*conscience collective*" and "*solidarite sociale*."

EVOLUTIONIST THINKERS

The idea of an "after-the-now" opened up a new historical perspective, and the search for what came before and how it developed characterized scholarship during the late eighteenth and nineteenth centuries. The "natural history method" became key to understanding a whole array of questions. "What was understood by this was the firm connection between what a thing is and how it got to be that way."[2] We enter an era when the "origins," "genesis," "advance," and "progress" of natural as well as social phenomena shaped most research projects. Between 1750 and 1840 a "steady stream of histories of various sciences poured from the presses, all purporting to demonstrate the existence and extent of progress in those disciplines."[3] The discovery of history as the key to understanding all natural and social phenomena was the "most revolutionary change of the nineteenth century." It was believed timely development, rather than "providential intervention," caused change.[4] It is difficult to find a significant thinker in the nineteenth century who did not work within the evolutionary paradigm.[5] This century, more than any other century, was self-consciously historical. Below I provide some examples.

Henry James Sumner Maine (1822–88) was a lawyer and expert in ancient Roman law. In *Ancient Law* (1861) he sought to explain the evolution of legal systems from "status" to "contract" in the Western world. He examined the development from kinship to territorial organization; civil to criminal law; and decree to case law and legislation. John Ferguson McLennan (1827–81) was another lawyer who carefully examined contemporary ethnographic records. In *Primitive Marriage* (1865) he postulated that there had been progress from matrilineality (descent through females), to polyandry (one wife, many husbands), and then to patrilineality (descent through males). In *Ancient Society* (1877), Lewis Henry Morgan (1818–81) also discussed the development of kinship and its relation to property and political organization. He believed that humans developed from *societas* (emphasis on person, clan) to *civitas* (emphasis on territory, property, state). Morgan also argued that kinship systems passed from matrilineality to patrilineality.

Herbert Spencer (1820–1903) is one of the most influential thinkers of the nineteenth century, and his ambitious encyclopedia, *Synthetic Philosophy* (composed of *First Principles*, *Principles of Biology*, *Principles of Psychology*, *Principles of Sociology*, and *Principles of Ethics*) was a thirty-six-year endeavor. He is remembered as a strong believer in social evolutionism and laissez-faire policies. Influenced by Comte, he attempted to show how society operated like an organism. However, he believed that society makes progress not just through the culmination of knowledge, but also through evolutionary principles. For Spencer, cosmic, biological, and social phenomena all evolve. As a system, society develops from "primitive," "ancient," to "modern" through different stages: (1) regulative system (which deals with the outside environment, providing defense and offense for the system); (2) sustaining system (economic production provides support for the regulative system); and (3) exchange and distribution system (communication and transportation networks and commerce and finance).

Edward Burnett Tylor (1832–1917), often called the "father of anthropology," believed in "psychic unity," or the idea that different peoples are equally equipped mentally and are therefore all capable of progress. According to Tylor, societies passed through savagery, barbarism, and civilization, and humankind's march through history represented a conquering of superstition by science. The faith of positivism shines through a passage from his *Primitive Culture* (1871): "Civilization may be looked upon as the general improvement of mankind by higher organization of the individual and society, to the end of promoting at once humankind's goodness, power, and happiness." James Frazer (1854–1941), who wrote one of the best known works in the Western world, the twelve-volume *The Golden Bough* (1890), is best known for popularizing anthropology. In his work he combined the most important intellectual concepts of the day: evolution, psychic unity, and progress. Humankind began its climb from animism and worked its way up through magic, religion, and finally to science.

Besides being the founder of modern Psychology, Wilhelm Wundt also wrote on what we might call cultural Psychology. He considered language, myth, and custom fundamental constituents of the "mind of a people" (*Volksseele*). His massive *Ethno-Cultural Psychology* (*Völkerpsychologie*, 1900–1920) is divided into ten volumes: The first two volumes focus on language; three on the arts; four through six on myth and religion; seven and eight on social organization; nine on law; and ten on culture and history. Wundt listed "four ages" of humankind's historical development: (1) primitive man; (2) the totemic age; (3) gods and heroes; and (4) the present age.

Other proponents of the evolutionary paradigm whose works are well known could be cited, but I will only mention them in passing. In *Capital* (1867), Marx discussed the development from equality to stratification. In *The Origin of the Family, Private Property, and the State* (1884), Engels discussed humankind's march from primitive communism's equality through feudalism, capitalistic stratification, and then on to a communist utopia. In *Division of Labor in Society* (1883), Durkheim wrote that European societies have progressed from "mechanical" solidarity to "organic" solidarity. In *Civilization and Its Discontents* (1930), Freud discussed our transformation from hedonists to people controlled by social restraints.

Though the reign of the evolutionist paradigm was seriously questioned (specifically by diffusionism, historical particularism, structural–functionalism, and certain developments in Psychology), influential intellectual continuities are apparent. For example, note the work of V. Gordon Childe (1892–1957) who, in *Man Makes Himself* (1951), emphasized the three great revolutions: agriculture, urbanization, and industrialization. Leslie White (1900–75), known for his *The Evolution of Culture* (1959), argued that societies are systems for harnessing thermodynamic energy, and that technology determines social structure, and social structure determines cultural forms. Julian Steward (1902–72) attempted to refine evolutionary thinking with his idea of "multilineal evolution," as seen in his *Evolution and Ecology* (1977), *Economic and Social Basis of Primitive Bands* (1936), and *Theory of Culture Change* (1955).[6] Marshall Sahlin's and E. R. Service's, *Evolution and Culture* (1960) is also relevant here. Marvin Harris, in *The Rise of Anthropological Theory* (1968) and other works, argues that the natural environment and technological level of a society determines its culture.[7] In addition, Roy Rappaport, in *Pigs for the Ancestors* (1968), attempted to demonstrate how a society (the Tsembaga Maring of New Guinea) is intimately related to the natural environment. These works may all be described as studies in "cultural ecology," or how the natural environment determines social organization and cultural forms over time. These approaches emphasize the natural–regulatory, not the moral–normative aspects, of a society.

Another scholar who was very concerned with "origins" was the Psychologist Julian Jaynes (1923–97). Though trained as a comparative Psychologist, he realized the limits of biological reductionism and sought clues to problems of the human psyche in history, literature, and religion. In *The Origin of Consciousness in the Breakdown of the Bicameral Mind* (1976; a key inspiration for this book), he argued that conscious subjective interiority was not a bioevolutionary phenomenon. Rather, interiority—and by this term he did not mean perceiving, thinking, or reasoning but the ability to introspect and engage in self-reflexivity—emerged historically as a cultural construction only about three millennia ago.

NOTES

1. Therborn, 1995, p. 4.
2. Robinson, 1995, p. 300.
3. Datson and Galison, 2007, p. 211.
4. Bayly, 2004, pp. 484, 315.
5. Though others offered counterarguments to the idea of "progress"—e.g., Ernst von Lasaulx, Nikolai Danilevsky, and Oswald Spengler.
6. See also Steward, 1953.
7. See also Harris, 1977, 1979.

Chapter Fifteen

Envisaging the Future as Paradise

We view the natural world with an unassailable "givenness," and science (derisively known as "scientism") is the prism though which we see "facts." The belief that such facts and facets of social existence—human nature, political organizations, and economic processes—were somehow rooted in "natural laws" has existed in the Western tradition since the ancient Greeks. However, it was not until the Enlightenment, the period that acted as the intellectual ideological soil from which the religion of progress grew, that it became acceptable to speculate that something called "society" could be objectively examined in the same way nature could be. This was revolutionary. It was then a short meta-framing step away from dissecting the social nature of humanity to transforming knowledge into useful technologies and applying scientific developments geared toward the refurbishing of society. In order to improve society and develop the necessary economic theories to do so, more concrete catechisms and scholarly disciplines would be required than those proffered by the early nineteenth-century utopians could envision. The tenets and practices of secular progressive humanism constitute training in the natural sciences and that exemplar of faith in universal truth, mathematics.

Building upon topics from the previous chapter, below I provide concrete examples of how the "promise of progress"—as a modern form of millenarianism—spread around the world, as well as how this revolutionary view of time implicated techno-science and rationalism.

THE PROMISE OF PROGRESS

It is important to point out that science, as a widely disseminated faith at least basically understood by the multitudes, is relatively new. Until recently, most societies were illiterate and lacked the educational experiences that indoctrinate faith in science and mathematics. In any case, between 1875 and 1900 in the Western world, an increasing number of research topics were viewed in terms of development. These studies tended to see things unfolding in an almost purposeful manner, or "less often, *growing* or *evolving* according to *laws.*" In most contexts "develop" meant "to unwrap or unfold," suggesting latency, potential, or some teleological promise about to be bestowed upon humankind (figure 15.1).[1]

TECHNO-SCIENCE AND RATIONALISM

The philosophy underlying and justifying science and technology is rationalism. This concept, which has become one of the most popular adjectives for describing how humans differ from animals, is usually associated with the Enlightenment (though it has historical precursors). We should note how rationality was considered an inherent trait of all cultures. Though some peoples may have scrambled up the ladder faster than others, eventually everyone would manifest rationality as a universal feature of the mature, autonomous, and fully interiorized mind.

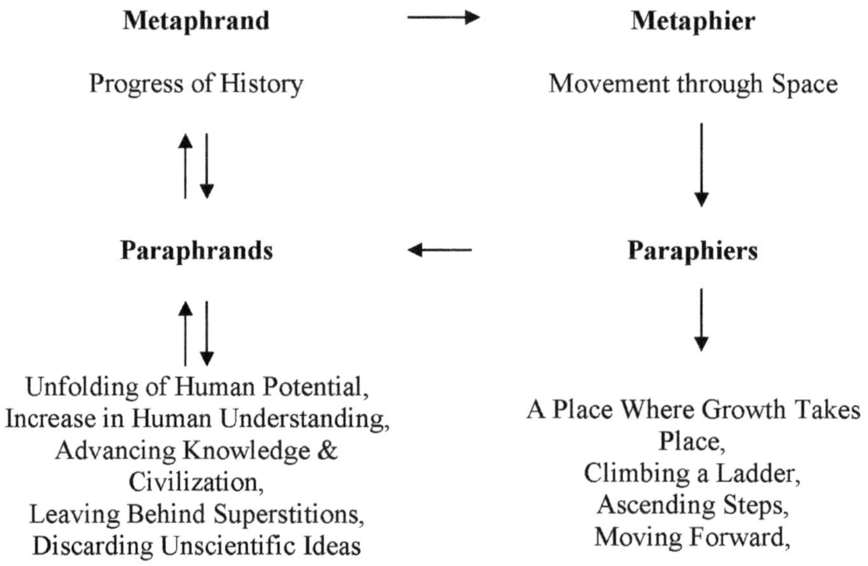

Figure 15.1. The Metaphoric Spatialization of History as Progress.

It was in nineteenth-century Europe that incessant technological innovation and rapid change in person–property regimes became a part of the intellectual landscape:

> The bourgeoisie cannot exist without constantly revolutionizing the instruments of production, and thereby the relations of productions, and with them the whole relations of society. Conservation of the old modes of production in unaltered form, was, on the contrary, the first condition of existence for all earlier industrial classes. Constant revolutionizing of production, uninterrupted disturbance of all social conditions, everlasting uncertainty and agitation distinguish the bourgeois epoch from all earlier ones. All fixed, fast-frozen relations, with their train of ancient and venerable prejudices and opinions, are swept away, all newly-formed ones become antiquated before they can ossify. All that is solid melts into air, all that is holy is profaned.[2]

VISTAS OF THE FUTURE

Since the 1800s, social theorizing has acknowledged the growth and complexity of human organizations, whether social, political, or economic. In the latter half of the twentieth century, this would give rise to "futurology" or "futuristics."[3] A fundamental difference, it should be noted, exists between futuristics and works of prophecy or the predictions of the likes of Nostradamus.[4] The latter concerns foreseeing events that transpire in a socially static world and are associated with premodern religious traditions and tracts, such as the Book of Revelation. But futurisitcs, rather than merely making predictions, describes, often in great detail, a world radically altered by scientific discoveries and technological developments that impact the political landscape and our place in it.

As Edward Cornish points out in *The Study of the Future* (1977), an emerging sense that a socially changed future is possible became evident in the late-eighteenth-century genre of speculative "prophetic fiction." For example, *The Reign of George VI, 1900–1925,* anonymously written in 1763, commented on the current state of society by describing future developments. However, society in the twentieth century was remarkably similar to that as seen in the 1700s. In 1770 Sebastien Mercier wrote *The Year 2440*, which described human civilization in AD 2440. It was published anonymously in Amsterdam since at the time it was taken as a criticism of the contemporary political system. In the future, peace reigns among nations and slavery has been abolished. However, this work did not speculate on how science and technology would greatly alter the human condition.

In 1852 Hans Christian Andersen wrote about tourists of the future. Jules Verne, who had a genuine interest in science, penned a scientific treatise about balloon navigation that was rejected. A friend suggested he fictionalize it into an adventure story. The result was *Five Weeks in a Balloon* (1863). He is also famous for *Twenty Thousand Leagues under the Sea* (1870). Significantly, the stories of Jules Verne, who authored about sixty books, were often grounded in solid science. Nevertheless, Jules Verne "peopled his fiction with nineteenth century characters living in a nineteenth century world; he did not undertake the far more complex task of trying to imagine how society as a whole would change as a result of the mechanical wonders he imagined."[5] H. G. Wells is well known for his fictional works, such as *The Time Machine* (1895). However, he also wrote nonfiction works, such as *Anticipations* (1901) that explored trends in society, and looked at issues of transportation, communication, urban growth, and military developments (in particular aviation). He explicitly saw a need for a science of the future.

In *The Coming Race* (1871) Bulwer Lytton described an advanced people called the Vril-ya who possessed domestic robots, television, devices that taught people while they slept, airplane-like machines, equality, and world peace. Unfortunately, because their civilization reached a saturation point, their instincts, ambitions, and energies declined, so that the Vril-ya became dissatisfied with their life. In Edward Bellamy's *Looking Backward 2000–1887* (1888), which can be called a "predictive utopian" work, future generations have radios, electric lights, full employment, and can retire at forty-five with pensions. And pollution does not mar existence (incidentally, Francis Bellamy [1855–1931] and author of the Pledge of Allegiance, shared his cousin Edward's utopianism). *In 100 Years* (1892), Charles Richet predicted that, based on demographic projections, by 1992 the United States and Russia would be the two most powerful nations. A work such as this must be seen within the context of the day. By the late nineteenth century the promises of science were no longer fantasy, and states began paying attention to and analyzing social issues using statistics for a variety of projects (especially for military purposes).

Pessimism has certainly been visible in religious writings throughout the centuries, so it should not be surprising that some did not believe the future would bless us with better times. Using science to make gloomy predictions can be traced to Thomas Malthus's (1766–1834) work on population and food. Unpredictability, apprehension, and dystopia became a staple of science fiction, especially during World War I when the horrific impact of advances in military technology displayed themselves with a bloody lethality never before seen on the battlefield. Pessimism is seen in H. G. Wells's *The Outline of History* (1920) and his *Mind at the End of Its Tether* (1945), as well as in the antiutopian nightmares that inform much of science fiction (e.g., Aldous Huxley's *Brave New World*, 1931) and political commentary (e.g., George Orwell's *1984*, 1948). Indeed, much of science fiction is characterized by dystopian themes.

CONCLUSION: THE LEGACY OF MILLENARIANISM

Various intellectual currents—e.g., unilinear time, social progress, economic growth, science and technology, national statism—eventually converged over the past several centuries to produce progressivism. This faith, colored by local cultures and specific sociopolitical needs, has many sects, denominations, and deifications that may be regarded as earthly incarnations, manifestations, and materializations of the Supreme Being of Progress. Over time faith in positivism has spread, both globally and within national states, from elite circles to the lower socioeconomic classes. However, because it eventually reached the point of being deeply held by so many, no formal dogma or church has been needed. Nevertheless, the prophets and priests of progress are still among us—optimistic techno-futurists, governmental plans for infrastructure development, profit and growth-hungry corporations, the self-help industry, and the ever-aspiring individual fed on consumerist promises of a better life just over the horizon (tbut note the exceptions of the Neo-Luddites and those who violently rail against "progress," e.g., the Unabomber Ted Kaczynski).

Why is belief in progressivism so strong? Many reasons could be offered, such as how it simply legitimates present politico–economic arrangements and ambitions. History has become more than just process. It is fulfillment. However, another reason is the confusion—among even the educated elite—of the difference between "natural evolution" and "social evolution." If the latter is a religion, the former is often grossly misunderstood. In any case, the two must be clearly differentiated. Belief in social evolution predates physical evolution, and it was Darwin, in his *Origin of Species* (1859), who so profoundly shook the world with notions about the adaptation of species. Despite the fact that many have glibly applied his thinking to sociocultural systems, Darwin himself was not a social evolutionist because his work was about biological evolution. Biological evolution occurs when physical characteristics that allow an organism to adapt to its environment are genetically transmitted. It does not mean that a species "improves" through time. It merely means species change. Society, of course, transmits knowledge via institutionalized learning, and an accumulation of knowledge certainly leads to change. But whether this change is "good" or not is a normative, not scientific, evaluation. To speak of "advanced" cultures is to make moral judgments, not factual statements. Additionally, to speak of "inevitable development" is to engage in mysticism, not science.

It must be pointed out that despite the assaults the religion of progress has suffered at the academic level, this faith is not seriously questioned, at least not in any profound sense, at the more popular level. Moreover, certain academic fields still accept the doctrine of inevitable human progress. Generally, some circles within political science, economics, and certain schools of sociology support and reproduce the doctrine of progress. Though their faith is sometimes muted and expressed in a sophisticated manner, it nevertheless shapes their research paradigm in subtle ways.

NOTES

1. Rosenthal, 1984, pp. 55, 40, emphasis in original.
2. Marx, 1954 (1848), p. 13.
3. See Edward Cornish, *The Study of the Future*, for a dated but useful treatment of futurology and "futuristics," 1977. Also, see Daniel Rosenberg and Susan Harding (eds.), *Histories of the Future*, 2005.
4. Harvey Cox, 1969. Cox notes three perspectives Christianity has about the future: (1) apocalyptic (negation; pessimism; retreat from earthly concerns); (2) teleological (optimistic; modern and secularized; purposefulness and direction); and (3) prophetic (the future is a realm of personal responsibility and hope; guarded optimism).
5. Cornish, 1977, p. 64.

Part IV

Self: Turning the World Inside Out

Chapter Sixteen

The Changeable Self through the Centuries

Rather than the needs and perspectives of anonymous masses and national societies, modern life is progressively structured around the "project of the individual." The "perspective of the individual, in one form or another, is increasingly celebrated."[1] This "institutionalized individualism" is not Western but operates at the worldwide level. The individual is the core element, the sovereign source of public life, whether political, economic, or cultural, as well as the source of problems in these areas; thus the modern emphasis on "therapeutic measures." The individual is the "primordial or grounding element of all social structures."[2] How did we get here? In this chapter I offer some broad illustrations in order to link conceptions of the individual, self, and conscious interiority.

THE VAGARIES OF CONSCIOUS INTERIORITY

Conscious interiority did not suddenly appear around the twelfth century BCE and then have its features set in stone, as if a light switch had been turned on. Conscious interiority is changeable, variable, and alterable. It is a historical process, not an all-or-nothing development that occurred only once. Its mutable nature is shaped by specific circumstances of a given period. For example, in *A World Made by Men: Cognition and Society, 400–1200* (1985), Charles Radding claims that changes in mentality were evident by the first centuries of the Roman Empire.[3] Indeed, as early as the third-century legal skills had already begun to decay. By the fourth century a mental world had emerged that "was already very different from that of classical times. Even occasional geniuses, such as Augustine, no longer fully understood classical authors." By the time of St. Benedict (circa 480–547) and his Benedictine Rule, the concern with distinguishing "thought from behavior had disappeared." It was "purity of deed more than purity of thought that monasteries offered secular society."[4] In other words, the duality of soul and body (though the division was not as sharp as the Cartesian dualism we moderns accept), premised on the privileging on interiority, had diminished relative to the classical period.

Below I explore the vagaries of what we moderns typically call the self by listing some key features of conscious interiority. Words listed immediately after the enumerated feature in parentheses are semantically related (and often extended); they are more prosaic terms that we commonly use in daily language.

- *Self-Authorization* (self-ownership, self-mastery, self-control): Our present inclination toward liberty at all costs is rooted in nineteenth-century views of "freedom" (even if it supposedly frays social relations and our connections to others). Politically, during the 1800s calls for independence, self-government, liberation, and nationalist movements in industrialized societies encouraged, psychologically, the idea that everyone has an "I" that should manage his or her "me." Aspirations for self-sufficiency echoed calls for more liberty in the realm of economics.
- *Self-Narratization* (self-help, self-improvement, self-development, self-authoring, self-directed, self-makeover): Self-authorization encourages individuals to envision themselves progressing on a linear temporality leading to future possibilities upon which one's "I" moves as protagonist. The heady and intoxicating atmosphere of rapid wealth accumulation, cooked up by the steam of factories, enticing window displays, and scintillating mass advertising, stressed immediate management of one's personal behavior. This reverberated with the idea of the "self-made man" (cf. the popularity of Samuel Smiles's 1859 *Self-Help* and Orison Swett Marden's 1894 *Pushing to the Front*). The value of reaching one's potentiality became, not just a societal issue, but also a psychologized reality. According to one observer, by the late twentieth century many people linked personal success stories with notions such as the "perfectibility of men" and "progress," thereby equating "mental health with the usually unreachable *ideal* state which combines success, love and lack of anxiety."[5]
- *Self-Autonomy* (self-reliant, self-made success): Resonating with self-authorization, political self-determination, and the promises of the self-made man, this feature describes how individual narratization bolsters a sense of self-control and a faith in one's destiny, thereby highlighting feelings of intentionality and responsibility for one's own actions and behavior. The difference between imputing behavior to the visible person or to an inner self is subtle and not always easy to discern. Nevertheless, history illus-

trates how understandings of the accountability for individual action—which is premised on the idea of self-autonomy—can shift from an external view (the observable person is held responsible) to an internal view (the "I" within the person) and back again depending on political economic forces.

- *Individuation* (self-discovery, self-actualization, self-realization, self-fulfillment, self-identity): In the nineteenth century one's "character" constituted one's identity and worth, but eventually it was one's "personality"—a more psychologized notion than character—that took center stage in understandings of the person. One's individualized self became differentiated and thereby came to be viewed as a collection of distinctive unique elements. When seen against the background of excerptions (another product of interiorization), one's personal traits were highlighted and contrasted in comparison with others.
- *Self-Reflexivity* (self-absorption, self-esteem, self-consciousness, self-awareness, self-fixation, self-glorification): "Seeing" one's self in an interiorized mind-space that lacks physical limitations, as well as the ability to imagine idealized incarnations of one's self, help produce an "I" that introspects upon a "me." The consequence is a "self observing self observing self" or an acute self-introspection that results in the mirroring effect. Such psychological processes can lead to obsessiveness as well as unrealistic opinions about one's abilities (i.e., the dark side of psychologization).

THE SELF-AUTONOMY AND INTENTIONALITY OF THE "INNER PERSON"

The Loss of Intentionality

What does it mean to claim that Europeans of the early Middle Ages "rarely concerned themselves with intention, whether they were assessing the mentality of an action or interpreting the words of authority"?[6] Or that "early Germanic society did not stress the importance of intention"?[7] Let us look at the evidence. But first some historical perspective.

In early Christianity, being a Christian was usually a personal decision, both internally (through repentance) and externally (through baptism). However, both of these experiences were not familiar to individuals of later centuries, since one was born into a society that had been thoroughly Christianized. Personal religious choice was not an issue. In the Germanic society that succeeded the Roman Empire, the system of "discipline had become strangely external."[8] What we have is a surprisingly "barren internal world,"[9] not the populating of psychoscapes with new experiences of which we moderns are more familiar. In other words, the center of the human condition was not the individual, and the experiences of his or her interior life were not privileged. In some respects, it seems as if certain experiences of interiority had been eroded. If

> we look comparatively at the relative, quantitative indications of self-examination and concern with the inner life of oneself and of others, it is easy to see that there was significantly more such interest in the twelfth century than in earlier medieval times. A quantitative graph, if one could be constructed accurately, would probably show a decline from the time of Augustine—roughly coincident with the Germanic invasions—only a slight increase in the Carolingian renaissance, a rise in the eleventh century which sharply increases in the twelfth, and then continuing though perhaps irregular increase up to the present.[10]

In terms of self-awareness and self-expression (i.e., the freedom of a person to state an opinion without too much concern about convention or the dictates of authority), the twelfth century was a remarkably inspired period.[11] The evidence suggests that the practice of self-examination was "deeper and more widespread in twelfth-century Europe than at any time since the fifth century."[12] Previously spirituality was formulaic, didactic, and inspirational; it was not about the person-as-a-unique individual with a particular relation to the divinity. It was around the twelfth century that the sense of community was relatively diminished, and new ritual practices came to signify not the "restoration of community," but rather the "kindling of personal devotion."[13] The individual has become more and more distanced—though not necessarily separated—from the community at large.[14] Concern switched from divine inspiration to personal expression,[15] and belief was becoming more important than behavior. Attention shifted from the salvation of all people to the deliverance of the individual.[16] The history of consciousness is characterized by increasing attention to a person's interiority.

Why was intention, one aspect of interiority that seems so central to our own definition of personhood, de-emphasized in Europe until approximately the twelfth century? After all, earlier Roman investigatory and judicial procedures were relatively sophisticated and fair by modern standards. Why were they replaced by practices that seem so unreasonable by classical as well as modern conventions?

The disintegration of a society's psycho-techniques involves many aspects, but one obvious example would be a loss of exposure to a multiplicity of views encouraged by literacy. Reading and an appreciation of the fine arts generate abstract, meta-framing thinking and imaginary spaces in which different stances vis-à-vis others can be taken. Another factor concerns the decline in urbanization. The lifestyle of the post-imperial period was decidedly more rural, so that experiences of social diversity—i.e., being forced to take another's interiority into account—decreased.[17] The deterioration of the city of Rome itself provides a sense of the profound cultural corrosion that shook the social order to its roots: Rome had perhaps half a million inhabitants, but the invasions had reduced its population to about fifty thousand. On top of this, its educated and cultured aristocracy had fled. The city would not revive until the Renaissance. Remember that justice by ordeal was accepted for centuries, and these pagan practices rested comfortably next to Christian beliefs. Indeed, priests often presided over the proceedings. Undoubtedly the political economic chaos caused by the Germanic invasions and the collapse of Roman civilization toppled sociopsychological foundations upon which understandings of intentional behavior rested. The weakening of intentionality as a key feature of interiority illustrates nicely the psychic plasticity of the human condition (rather than psychic unity).

Consider confession, which after the early Middle Ages became more important as individuals relied less upon the hierarchy of the Church and priestly absolution and more on their own personal sincerity. An "awareness that external penance was distinct from inner repentance, and that it was the latter which God valued" became more explicit.[18] In other words, a type of dualism was emerging that gave increasing significance to interiority. For us moderns, guilt is tightly linked to intention, and for jurists of the late Roman Empire and certain Church fathers, it was a matter of legal and moral concern. However, after Germanic culture eclipsed classical civilization, intention lost its significance.[19] Eventually, the examination of one's conscience would again become associated with confession and intention. To use an old staple from the anthropological literature, societies evolved from a "shame culture" to a "guilt culture" in other words, guilt is interiorized shame.

Also consider Germanic law between the fifth and the twelfth centuries, which "rarely tried to take into account the intention of an actor when assessing punishment"; it did not make a clear-cut distinction between wrongs committed intentionally or unintentionally. For example, Salic Law, codified by the leader of the Franks, Clovis (481–511), had little to say concerning intention.[20] The "penal codes usually prescribed punishments for actions and not for the intentions behind them, and the same was true of the Penitentials, where penances were attached to external acts rather than to internal states of mind" [interiority].[21] Attributing guilt was highly ritualistic: Individual accountability was determined by "ordeals" (from the Germanic term meaning "judgment") which revealed "judgments of God" (*judicia dei*). For example, in the water ordeal the accused would be bound and thrown into a river. If the person floated, the verdict was innocence; if he or she sank, the accused was guilty. Another type of ordeal was by fire: the accused would reach into a pot of boiling water to retrieve a stone or hold a red-hot iron. Quick healing demonstrated innocence, or the "burned areas were sealed for three days and the case was lost if there were signs of infection when the bandages were removed." Judicial duels or ordeals by combat were another way of discerning God's judgment: the winner was innocent and the loser guilty. Finally, compurgatory ordeals, in which a defendant typically would swear a prescribed and strictly worded oath as to the veracity of his or her statement, were employed. Then a number of compurgators would be summoned who also took the oath. However, those who swore an oath were not character witnesses in the modern sense. The oaths they took were ritualized, verbal formulas that needed to be repeated perfectly for the oath to be effective. [22] As further evidence that intentional interiority of the individual did not matter, substitutes could stand in for the accused and accusers.

The Reinvention of Intention

In Europe by around the twelfth century fundamental changes were afoot in how individuals conceived of their relation to the cosmos. Why did this shift occur? Urbanization, more trade and commerce, increased circulation of wealth, the revival of classical texts, and better education all played their part. However, such changes in themselves cannot explain this transition, since the makeover in mentality itself encouraged by these political economic transformations must be recognized. As evidence that individual interiority was now being privileged, consider the fact that since courts began to regard mental states as carrying significance, new doctrines of law were created. It was also during this period that the practice of excusing children and the insane from their actions began. Furthermore, the concept of chance starts to supplant divine judgment so that individuals begin to doubt that all events were imbued with supernatural meaning.[23]

The "reconceptualization of the relation between the human mind and the world of things was probably the most revolutionary product of the new cognition of the high Middle Ages, because there is to be seen a break not only with the early Middle Ages," but with classical antiquity as well. Changes in cognition are apparent in not only what scholars of the time investigated, but also in *how* they studied certain topics. It was around this time that Medieval philosophers discussed "intention," from the Latin *intendo*, meaning to point at, aim at, or extend toward something. And consider the Medieval disputations about conceptualism, nominalism, and realism. The pertinent philosophical technicalities need not distract us, except to say that these debates concerned what is "within the person" and "what is external." For centuries, beginning with Aristotle, it was believed that "a union of the senses with the source of sensations" occurred ("conjunction"). By the twelfth century, however, this view was being challenged, and interior processes increasingly became cut off from external reality. Advocates of realism (e.g., Anselm, Archbishop of Canterbury, 1033–1109) and conceptualism (e.g., Peter Abélard, 1079–1142; associated with Scholasticism and one of the most critical thinkers for his time) deeply pondered the apparent disjunction between thought and reality. They set the groundwork for the very modern notion that an ontological dichotomy exists between knower and what is known; mental imagery became regarded as imaginary or fictitious. In other words, due to the changing social environment, they turned their attention to a new introcosm. The crucial point is that what changed was not just doctrinal belief but mentality itself.[24]

PERSONAL UNIQUENESS AND HEIGHTENED INDIVIDUALITY

Privileging Individual Traits

Concomitant with increased intention is the explicit attribution of individuality. The concept of the "individual" (from "indivisible") has been a central assumption of the human condition that has underpinned much of Western thought (i.e., "individualism" in its broadest sense).[25] However, we should not assume that there is anything inherently "individual" about the biologically singular person; it is the historical contingencies of political philosophy and economic arrangements that have shaped our view that the person is not divisible. Let us look at the ancient Western world in order to gain some perspective on the "individual."

The Emergence of Individuality

A "great leap in subjectivity separates" Medieval Europe from the eighteenth century.[26] In the Medieval period, discussions of the person as an inner-oriented psychologizing unit did not rely on abstract notions such as "individual"; terms such as *individuum, individualis,* or *singularis* belonged to "logic rather than to human relations." Rather, more concrete expressions, such as "knowing oneself" or "considering oneself," were employed. The word that was perhaps closest to our "mind" was *anima,* which really meant a spiritual entity (soul). Also the spatially oriented "descending into oneself," "inner man," or "inner mystery" were sometimes employed.[27]

We should not confuse "self-awareness" with "individuality." The former does not necessarily entail individuality, while the latter implies the uniqueness or special traits of an individual. In premodern times it was not unique personality traits, but exemplary models that were admired; not how to be one's self, but how to be someone else's self. The Homeric hero, the Germanic hero, the "polis-minded" man, the Roman *pater familias,* Aristotle's "great-minded man," the "unshakable Stoic," the knight, the gentleman, and the teacher, prescribed certain virtues, values, and traits. They offered a "script" for life.[28] It is worth noting that, at least in Europe, the Medieval person "could never verbalize the idea of having a 'personality.'" Indeed, no word with the meaning of "personality" existed, and during the twelfth century "persona" primarily still meant a mask held before an actor's face.[29] Also, around 1000, "artists were little, if at all, concerned with the recording of individual likeness."[30] A significant difference, then, between "portraits," and "personal portraits" can be seen; the latter is a representation sensitive to the individual as an interiorized being with unique traits. By the 1300s French portraits began to look like "specific people, rather than generalized types." These became fashionable as book illustrations.[31] In modern times "we are captivated by an uncanny sense that each one of us constitutes one irreplaceable human form, and we perceive a noble life task in the cultivation of our individuality, our ineffable self." Moreover, it is worth noting that autobiography, at least its modern form, is not older than the eighteenth century.[32]

In the writings of the Medieval period can be detected a proto-Psychology, though the interest in these texts was to "study the Godward movement of the soul the things which impeded and encouraged it." It was a highly spiritualized Psychology, or alternatively, a "clinical theology." After all, our purpose in life was not to be human, but to become angelic; our goal was not to realize our true selves, but rather to be transformed into something quite different.[33] This may be difficult for us moderns to appreciate, but remember that the period in Western history under consideration was one of world-renunciation, not world-affirmation; escape from this world, not a welcoming of its joys and beauty.

From where did the "individual" come?

> The emergence of notions of individuality, in the modern sense, can be related to the breakup of the medieval social, economic and religious order. In the general movement against feudalism there was a new stress on a man's personal existence over and above his place or function in a rigid hierarchical society. There was a related stress, in Protestantism, on a man's direct and individual relation to God, as opposed to this relation mediated by the Church.[34]

During the late sixteenth and seventeenth centuries, "interest in history, biography, and exploration testified to the expansion and liberation of the human personality [interiority] which was the dominant characteristic of the period."[35] From the 1600s, "individual" came to mean a single distinguishable person. It could have pejorative associations, denoting idiosyncrasy, eccentricity, or vanity. It would eventually be used in early biology to denote the "ultimate indivisible division" of a species. It began to be used in its modern sense in the 1700s, but it was not until the 1800s that it gained common currency.[36]

Processes of Individuation

Before people assign great value to interiority, they must first regard the person as an individual—rather than an individual's role or social position—as inherently valuable. This process, which can only be appreciated by adopting a long-term historical perspective, can be called individuation (or individualization). This describes the emergence during the last several centuries of a profound concern with being methodically numbered, listed, catalogued, uniformed, examined, and measured.[37] These practices isolate and identify the individual from the amorphous social mass for purposes of political economic organization and mobilization. Individuation, then, concerns how the forces of modernity (e.g., statism, industrialism, capitalism, science, new technologies) recorded the behaviors of the individual, who were turned into diligent students, productive laborers, and loyal citizens. Individuation encourages the thinking of the person as a unit in possession of a personalized and unique interiority. Conscious interiorization has been accelerating since the 1800s due to political economic conditions that position the individual at the center of society. Additionally, scientific advances encourage us to "see" reality via mediated technologies, thereby strengthening our mind's eye to introceive.

The processes of individuation—being enumerated, grouped, examined, and observed—are referred to as "disciplines" by the French philosopher Michel Foucault. The principles of discipline are: (1) spatialization (a place for everyone and everyone in his or her place); (2) minute control of activity (increasing control of bodily behavior); (3) repetitive exercises (to instill the proper habits and automatic responses); (4) detailed hierarchies (each level keeps watch over the lower ranks); and (5) normalizing judgments (which keep one from deviating from the official line).[38] It is worth mentioning that in general, the least powerful become the most individualized. One thinks of soldiers, convicts, prisoners, and perhaps, to extend the argument, students. We can extend the argument even further, and note that, in a certain sense, women are more individualized than men. After all, it certainly seems that in many modern cultures the latter are expected, more than men, to exercise bodily control, dress up, and decorate and enhance their appearance with jewelry and cosmetics.

In our own times, one manifestation of individuation is the great attention given to bodily control, self-monitoring, and hygiene. These are required to demonstrate the degree of a person's integration into the productivist political economic regime. Note that compared to people from the Medieval period, our concern with bodily cleanliness (and cleanliness in general) would seem bizarre, eccentric, and inexplicable.[39]

Besides the "external" dimension of individuation, another aspect of this process concerns how, as a type of emergent subjectivity, it is constructed from the "inside out." Phrased differently, the more political economic institutions shape an individual's observable behavior, the more interiority emerges as a response to outside individualizing pressures. Individuation, then, increases psychological interiorization and encourages us to view individuals as containing mental stuff. The individual's interior space generated by individuation not only requires expression but is also filled with very personal demands. The forces of individuation dug the groundwork for individualism and the invention of definitions of the person that crystallized in nineteenth-century Europe. These include the:

- Voting citizen: Belief in the inherently autonomous self.
- Participatory citizen: Units for mobilization demanded by national states.
- Consuming citizen: Property-accumulating units with inalienable economic rights.
- "Subject": The focus of research Psychology; a container of unconscious impulses.
- "Character": The self-inspecting protagonist of the novel.

What these different understandings of the person share is the notion of the person possessed of a vast interiorized space that requires expression and exploration, and they illustrate how individuation and individuality are cut from the same ideological cloth.

NOTES

1. Meyer, 2004, p. 6. See also Meyer, 1986.
2. Frank et al., 1995, p. 360.
3. Radding adopts a Piagetian approach and explores changes in the "reasoning" of individuals. However, he stresses that his use of "mentality" should not be taken to mean that "society as a whole has 'mind' that is capable of development of evolution." Also, the fact that one person reasons at a cognitively more advanced stage than another does not "prove that the first is a morally better person or a wiser one than the second" (Radding, 1985, p. 33).
4. Radding, 1985, pp. 76, 73, 105, 108.
5. Gross, 1978, pp. 9, 5 (emphasis in original).
6. Radding, 1985, p. 153.
7. Morris, 1972, p. 74.
8. Morris, 1972, pp. 24, 70.
9. Benton, 1982, p. 263.
10. Benton, 1982, p. 288.
11. Morris, 1972, p. 7. Between the millennium of 500 to 1500, Morris claims the West witnessed "three renaissances": Carolingian, the twelfth century, and the fifteenth century.
12. Benton, 1982, p. 264.
13. Morris, 1972, p. 12.
14. Ong, 1982, p. 178.
15. Benton, 1982.
16. Morris, 1972, p. 152.
17. Radding, 1985, pp. 5, 29.

18. Morris, 1972, pp. 73, 71.
19. Benton, 1982, p. 272.
20. Radding, 1985, pp. 92, 93.
21. Morris, 1972, p. 74.
22. Morris, 1972, p. 74.
23. Radding, 1985, pp. 242, 241, 32.
24. Radding, 1985, pp. 210, 158, 207, 251.
25. It might be noted that Medieval theological debates about the unity of the Trinity relied on an idiom of what was "indivisible." More narrowly defined, individualism is the Anglo–American political economic philosophy adopted and promoted by much of the Anglophone world.
26. Benton, 1982, p. 266.
27. Morris, 1972, pp. 64, 65.
28. Weintraub, 1978, p. xv.
29. Benton, 1982, p. 284.
30. Morris, 1972, p. 89.
31. Crosby, 1997, p. 166.
32. Weintraub, 1978, pp. xiii, 26.
33. Morris, 1972, pp. 76, 31.
34. Williams, 1983, pp. 163–64.
35. Palm, 1936, p. 97.
36. Williams, 1983, pp. 162–63.
37. See Shilling, who defines individuation as a "set of practices by which individuals are identified and separated by marks, numbers, signs and codes which are derived from knowledge of the population and related to the establishment of norms" (1994, p. 78). See also Turner, 1983.
38. Foucault, 1979.
39. Cf. Elias, 1978, 1982.

Chapter Seventeen

The Narratized Individual as Social Actor

In the Middle Ages a self-conscious concern with individuality and the need for "complex processes of self-exploration or self-orientation" were not salient. Indeed, bringing attention to one's individuality might have even been an "embarrassment."[1] When authors did write about themselves, they were motivated to offer *exempla* of their spiritual struggle for their reading audience. The point is subtle but significant: A difference exists between being one's own biographer and doing autobiography. The latter is a modern genre, while the former was not an attempt to gain greater objectivity, but rather to make the author into an actor fulfilling a socially sanctioned role.[2] The point here is that the more individuals possesses a sense of ineffable individuality, the more they will, in moments of self-reflexive decision-making, ask themselves what decision would genuinely express my special character? "And the less will he find comfort in models."[3] For example, consider the irascible and ambitious Bishop Ratherius of Verona (887–974), whose tumultuous life, one would imagine, should result in some very introspective writing. Though a sense of self-reflection in some of his works is evident (he did, after all, criticize his own actions), Ratherius

> does not present a view of his own self as a writer would do who is conscious of an inner organizing center in the personality. He instead describes himself as a person whose inner order is dependent on the fact that there is an order in the surrounding world in which he exists. He knows himself neither as a person who projects the order of his own experience on a surrounding world (as some modern men conceive of this) nor as a self-regulating organism coexisting with a surrounding world having its independent order.[4]

Consider another example from a later period, James I (1208 or 1213–76), the king of Aragon and one of the greatest soldiers of the time. His autobiographical *On the Deeds of James* possesses "that 'naïve' quality in which the recollection of the act has not been greatly transformed by conscious reflection on it." In other words, the "mental distance between actor ["I"] and recorder ["me"]" cannot have been great.[5]

How can these premodern views of the person, so different from our self-centric worldview, be explained? In this chapter I contextualize the meaning of the "inner person" by examining the historical vagaries of individuality. How definitions of the person changed—becoming more psychologized—are discussed by investigating the emergence of its newer, more modern aspects—e.g., self-autonomy, individuation, self-reflexivity, and personal uniqueness.

A LINEAR TEMPORALITY WITH THE SELF AS MAIN ACTOR

Jaynes points out that within each of our individual introcosms we are always "seeing" ourselves as the central characters in the drama of our lives. However, a certain selectivity to this inner theater is apparent, since we choose those images of ourselves that fit into our favored scripts and ignore those that do not. This attribution of causes to behavior or "saying why we did a particular thing is all a part of narratization. Such causes, as reasons, may be true or false, neutral or ideal." An interiorized self is "ever ready to explain anything we happen to find ourselves doing," and a "stray fact is narratized to fit with some other stray fact."[6] We are not "consciously aware of all the information our mind processes or of the causes of all the behaviors we enact, or of the origin of all the feelings we experience. But the conscious self uses these as data points to construct and maintain a coherent story, our personal story, our subjective sense of self."[7]

The ability to narrate rests upon that most basic feature of interiority, spatialization. Narratization indicates an increasing focus on the inner person. Among other abilities, spatialization allows us to mentally envision an imaginary trail with a beginning and an end upon which the "I" stands, introspecting (looking at the present "me"), retrospecting (looking back at earlier "me's"), or prospecting (looking forward to "me's" yet to be). Narratization, then, impacts how individuals conceive temporality and their movement through it. However, if a society for whatever reason, decides to emphasize the narratizing feature of interiority at the macrolevel, then history becomes no longer cyclical and the present is no longer measured only by what came before. This is indeed what happened in late-eighteenth-century Europe. A new perspective on time, now conceived in linear terms, profoundly transformed the way we view the world. Cosmology, scientific understandings, and the purpose of political economic institutions were all impacted by one of the most momentous changes in global history: the invention of the future.

AUTOBIOGRAPHY AS SELF-REFLEXIVITY AND SELF-NARRATING

Certain features of interiority that we take for granted—indeed, that we highly privilege, such as individuality or sustained self-scrutiny—are lacking in the historical record (or at least not given much importance). Though individuality is more clearly delineated in the twelfth-century biography than it is in autobiography (suggesting, counterintuitively, that it is easier to describe another person than ourselves), the premodern Western autobiography is not autobiographical in the modern sense.[8] In the Western tradition, the "full convergence of all the factors constituting this modern view of the self"—self-awareness and individuality of the person—emerged "only at the end of the eighteenth century." Thus, what we would consider genuine autobiography is much rarer before 1800 than we might expect.[9]

Ancients certainly wrote about the great deeds they accomplished (*res gestae*), memorable events (*memoir*), and their lives. However, as Karl Weintraub makes clear in *The Value of the Individual: Self and Circumstance in Autobiography* (1978), "none opened up their souls in the inwardness of genuine autobiography." Speaking of the Emperor Augustus (27 BCE–AD 14) and his *Res gestae*, Weintraub notes that "one of mankind's most extraordinary personalities records his deeds and the intent of his policies—but reveals absolutely nothing of his inner life." In the Emperor Marcus Aurelius's (121–180) *To Myself*, there is similarly "no active self-exploration," no self-revelation "for the profit of others, no real self-portraiture":[10]

> During the millennium from 800 BC until 200 AD, the conditions of ancient life neither stimulated nor promoted the growth of autobiography. The ancients did not put a premium on the life devoted to settling the quandary: who am I? How did I become what I am? In what sense am I a distinctive personality? And what interplay of external forces and internal characteristics accounts for my specific configuration?[11]

However, toward the end of the classical world there "suddenly looms that book whose sheer presence makes one feel like the traveler through the plains who suddenly comes upon the mountains."[12] It is striking that, at least in the Western tradition, until "the beginning of the twelfth century no reader of the *Confessions* dared or was moved to write a self-examination in the same mode." Indeed, it must be noted that in his concern with self-inspection St. Augustine is a "thoroughly unrepresentative figure of his own period, a mountain, as it were, surrounded only by hills."[13] Though authors of the Middle Ages would look up to the *Confessions* as a model, they did not achieve the "inner unity that so characterizes Augustine's work. None has the "scope, the fullness, the inner richness, and the intense personal focus" of his *Confessions.* The nature of time and memory, key aspects of the narrating self, play a salient role, so that this book approaches what we might call genuine autobiography. Nevertheless, we must note that even St. Augustine was not interested in unique traits of the individual or personal idiosyncrasies. These apparently had no inherent value. Rather his purpose was to offer his own experiences as a typical life to be followed for religiously didactic reasons.[14]

Visualizing the Narratized Individual

In the visual arts "a similar decline and then reemergence of interest in the individual are apparent" from the classical to the late medieval period. Between the end of the classical period and the late Middle Ages, the rank and status of the individual were often evident in portraiture, but personal features were not. However, between about 1000 and 1260 "a movement from a portraiture which concentrated on hierarchy and stations, to one which was keenly aware, at least on occasion, of personal appearance" was evident. Though one observer writes that twelfth-century Europe produced very few individualized portraiture,[15] it does appear that a "personalization of the portrait" occurred after 1000 and the more and more personal details in portraits became discernible.[16]

Somewhat later, a key the feature of interiority, narratization, would emerge in the arts. "European authors in the twelfth century had a clearer sense of their own inner life, and their relations to others than their Carolingian predecessors. A heightened sense of history is a form of self-consciousness."[17] For instance, consider a work by the brilliant Florentine lyrist, Francesco Petrarch (1304–1374), commonly known as the father of Renaissance humanism. In his *Secret Conflict of My Cares* (*De secreto conflictu curarum mearum*, 1342), three dialogues are described between St. Augustine and Petrarch himself, while "Truth" sits in the corner of the room as "the guarantor of an unrelenting dedication to veracity." Petrarch, in an example of meta-framing, created a scene in which he ("I") talks to himself ("me") about Petrarch (that is, his other "me's"). Interior spatialization and the narrating self allow the "Petrarch that was [past "me's"], the Petrarch that might have been [hypothetical "me's'], and the Petrarch that might be" [future "me's"].

This new enlargement of subjective sceneries with different selves, temporally positioned within mental space, can be described as psychological perspectivism, or a "procedure of introspection" in which, in a meta-framing maneuver, one steps outside of oneself.[18] One is tempted to suggest that the inner perspectivism of the person apparent in Petrarch's work foreshadowed the use of perspective in the visual arts, or the science of painting in which objects display depth and distance. Perspectivism, after all, developed about a quarter century after Petrarch's death. In other words, in the same way that increased interiority led to a mental perspective in which one's "I" could view one's "me's," perceptions of the external physical world underwent an analogous change.

NOTES

1. Weintraub, 1978, pp. 51, 71.
2. Benton, 1982, p. 266.
3. Weintraub, 1978, p. 89.
4. Weintraub, 1978, pp. 56–57.
5. Weintraub, 1978, p. 66.
6. Jaynes, 1976, p. 64.
7. LeDoux, 1985, p. 206.
8. Benton, 1982, p. 269.
9. Weintraub, 1978, pp. xv, xiv.
10. Weintraub, 1978, pp. 1, 16–17.
11. Weintraub, 1978, p. 13.
12. Weintraub, 1978, p. 1.
13. Benton, 1982, pp. 265, 288.
14. Weintraub, 1978, pp. 91, 1, 24, 45.
15. Benton, 1982, pp. 274, 288.
16. Morris, 1972, pp. 87–90.
17. Benton, 1982, p. 285.
18. Weintraub, 1978, p. 98.

Chapter Eighteen

The Self as Mirror in Historical Perspective

What is the "self"? Much of the literature on the "self" is dissatisfying since very few ask what exactly is the "self." Even though the self is "one of [P]sychology's central concerns, the self rarely gets directly defined by those concerned with it"[1]: Is it the perceiving agent (the subject that is aware)? Or what is perceived (an object of awareness)? The truth is that the self is both subject and object. Indeed, because its nature has not been adequately theorized, the result has been a mystification of the self. It becomes a black box. Rather than assuming the existence of irreducible, metaphysical atom-like selves, we should regard selves as being divided and subdivided by social forces, i.e., into different personas and roles, depending on the social circumstances. Though we should inquire into what ways aspects of the self are universal, we should also ask if the self can be further broken down into constituent parts. In the same way that physicists must rely on a type of radical methodology—highly abstract, meta-framed theorizing apparently disconnected from the visible world, measurements of indirect phenomena, and supersensitive instruments—those of us interested in the nature of the psyche should also become radical methodologists and move beyond commonsensical empiricism. In the same manner, then, that subatomic particles can only be detected and studied indirectly, so the component parts of the psyche must sometimes be researched in a roundabout way. The self needs to be, in the same manner the atom was broken down into electrons, protons, and neutrons, taken apart into its constituent elements.

This chapter continues the discussion of self-reflexivity from the perspective of how the analog "I" and the metaphor "me" relate to each other. This analysis links up with how the diversification of roles, due to an increasing complex political economic landscape over time, has enhanced psychological interiorization.

THE ANALOG "I" AND THE METAPHOR "ME"

I begin with the premise that enough evidence has been accumulated to demonstrate that two aspects of the self—i.e., an active "I" and passive "me"—exist (in linguistics, the dramatic arts, and religio-philosophical thought). This bifurcation of self (conveniently described as a distancing between an "I" and a "me") reflects the needs of society to construct an analog microcosmic society within the person. In other words the "external" aspect of the person as an observable doer or receiver of action is "miniaturized" into an "internal," subjective, inner sense of being an agent ("I") or recipient ("me").[2] This miniaturization of society is not inherent to the human condition, but rather results from the way social organization shapes the psyche due to increasing complexity and role differentiation. Internalized social relations, then, become "a society of selves" with the analog "I" and a metaphor "me." A role, after all, can be described as "the capacity of being several things at once."[3]

What of this split self? The analog "I" is the active subject, the "doer" that initiates action. It can "'move about' vicariously in our 'imagination,' 'doing' things that we are not actually doing." This allows a host of cognitive abilities, such as imagining possible outcomes after making decisions, mentally picturing ourselves doing things that we are not carrying out in the physical–behavioral world, and long-term deception. The "I" allows "autoscopic images" within our personal introcosm: "We can both look out from within the imagined self at the imagined vistas, or we can step back a bit and see ourselves perhaps kneeling down for a drink of water at a particular brook."[4] If the "I" is an active subject, carrying out actions, then the "me," as passive object, receives actions in our mental space. Examples of the passiveness the "me" experiences include hypnosis, spirit possession, and New Age channeling. If the "I" feels in control, the "me" feels controlled. Together the "I" and "me" constitute the self, and as a form of cognition, the "I–me" dynamic may be either "deep" (held with strong conviction and difficult to change) or "surface" (held with little conviction and easily changed).

During everyday behavior, the "I" and "me" are usually equated with each other. Or to quote Goffman, "The performer comes to be his own audience; he comes to be performer and observer of the same show."[5] However, when one is not being oneself—e.g., acting, lying, disassembling—the "I" and object "me" are separated because the former is carefully supervising the latter.[6] In other words, when we deceive others, the "I" (true intentions) separates from the "me" (dishonest performance in front of others). When we act on stage, the "I" (actor) directs the "me" (character).[7] When we self-deceive, the "I" and "me" are so overly identified with each other that a meta-framed space for critical reality checking dissolves.

As Jaynes pointed out, "quite profound problems" regarding the relation between the "I" and "me" confront us, but for our present purposes we should note how the "I–me" dynamic leads to enhanced self-reflexiveness and self-referentiality. It is these "vicarial selves" of "I's" and "me's" that populate our psychoscapes, resulting in the fractal complexity of self-reflexivity.[8] The more one's mind becomes introspectable, the more opportunities arise to "see" one's self as an entity exploring ever-newer psychoscapes. Over the centuries a sense of control was felt to "arise more and more from *within* the individual" (emphasis in original of nature). This increasing attention to "inner processes" corresponded with the trend to view natural animistic forces immanent in the world as less salient. What was external to the individual (e.g., impersonal laws) became more significant. The world was "turned inside out." This sentiment, "an obvious and early fact of experience to every one of us" and a "fundamental starting-point of our life as conscious beings," is actually "quite a recent achievement of the human spirit" (from around 1000 BCE). Over time the distinction between "my self" and all other selves—or if phrased differently, the antithesis between "myself" (the observer) and the external world (the observed)—became sharper.[9]

SELF-REFLEXIVITY: THE "I" MIRRORING THE "ME"

It is absolutely essential to distinguish between perception of the physical world versus introception of one's mental landscape. Not only does the average layperson confuse these crucially different types of cognition, but so do not a few Psychologists and philosophers. To see the difference, as well as to appreciate the meaning of self-reflexivity, take the following behavior as an example: I see myself (perhaps in a mirror or, like Narcissus, in a calm pool of water). This cognitive event in and of itself is not a subjective, self-aware process. It is complex cognition, to be sure, but such mentation does not require conscious interiority.[10] Now let us meta-frame this act of thinking by interiorizing it (indicated by putting mentally spatialized entities in quotations): "I" "see" "me"—i.e., while looking at myself, I become conscious of the very act of observing myself. We can climb the ladder of abstraction by meta-reframing even further, thereby generating recursive self-reflexivity in a series of stages or orders (quotations designate mentally interiorized behavior):

- First Order: Perception = I see myself
- Second Order: Interiorized Cognition = I "see" ("I" "see" "me")

- Third Order: Self-Reflexivity = I “see” ([“I” “see” “me”] “see” [“I” “see” “me”])

The transition from perceptual to interiorized cognition may be hard to grasp. However, the change from stage two to three is even more difficult to explain, but let us introduce a term for it: the mirror effect. This is a result of meta-framing. It occurs when the “I” (observing subject) and “me” (observed object) switch roles (as if, when viewing ourselves in a mirror, we become confused about the difference between the physical me and the reflection). This role switch—the “I” becoming “me” and vice versa—is possible because both the “I” and “me” share personal attributes of the individual, thereby reflecting to each other the same traits of one’s selfness (in figure 18.1 note that “myself” is written as “me”).

THE REFLEXIVE SELF EMERGES

The explosion of self-reflexivity in modern times is evident in ways major and mundane. A few words on the emergence of self-reflexivity in the Anglophone tradition will provide some historical perspective. Originally in Medieval English, “self” merely emphasized or reflected back on or indicated identity. Eventually, a need arose to speak of the act of reflecting on one’s self, and somewhat later in the Renaissance, reflexiveness became a “subject of investigation, of celebration, of controversy.” However, there was still no “self” in the modern Psychological sense. “Soul” or “man” was used instead. This was, after all, a pre-Psychological period.[11] During the seventeenth and eighteenth centuries self did not possess a nominal definition, but rather carried an emphatic or reflexive meaning in the purely grammatical sense.

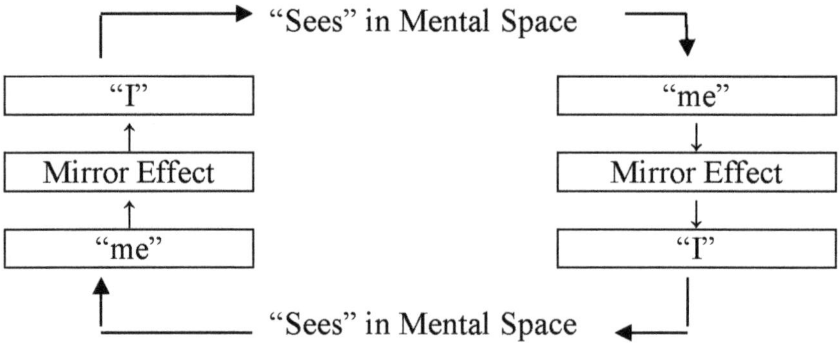

Figure 18.1. The Psychodynamics of Self-Reflexivity.

Eventually three rather different definitions of "self" emerged. The first is the "self as self-interest." This was the earliest meaning, and it was obviously negative, denoting selfishness. "Selfish" appeared in 1640. "Self is the great Anti-Christ and Anti-God in the World" according to the 1680 Oxford English Dictionary. Terms such as self-praise (1549), self-love (1563), self-pride (1586), self-contained (1591), and self-regard (1595) expressed these negative sentiments. The second definition is the self as a "principle of unified identity." This described a permanent subject of reflexive thought, and such a definition became clear in seventeenth-century philosophy and was used to describe the individual as a thinking, conscious being. This is how John Locke used "self," but as Rosenthal points out, he also used "person" for the same purpose, "a sign of how problematic this definition of self was."[12] The seventeenth century saw the coining of self-pity (1621), self-knowledge (1621), self-denial (1642), self-interest (1649), self-confidence (1653), self-esteem (1657), self-created (1677), and self-determination (1683; the political sense of the latter term did not appear until 1911). Later even more new self concepts emerged, seen in terms such as self-control (1711), self-possession (1745; meaning "command of one's emotions"), self-important (1775), self-justification (1775), and self-respect (1795). An English dictionary might list 140 to 150 words prefixed with "self-."

The third definition is the self as an "element of divided identity" or the partial self. This final usage conflicts with the sense of permanency, as in "what one is at a particular time" (e.g., "my former self" or "my later self"). And "what one is in part" (e.g., "my better self," "natural self," or "spiritual self"). This divided self is capable of changing and linked to notions of progress and self-improvement. Today we use all three senses of self, but as Rosenthal points out, it would not be until the twentieth century that a truly positive value was assigned to the self. We can label this new formation as "Psychology's—interiorized—self":

> The new internal psychic world in which agency and emotion are relocated is that of the "self", and this world is, in its turn, also increasingly subject to the cool, dispassionate and inquiring gaze which disenchanted the outer, with the result that consciousness of "the world" was an object separate from man the observer, was matched by a growing consciousness of "the self" as an object in its own right.[13]

Moreover, self-reflexivity, a type of meta-framing, encourages us to mull over our subjectivity in new ways. "By separating anger from aggression, or fear from flight, a start is made on the process by which emotion becomes defined as a largely interiorized facet of human experience."[14] And long-term developments demonstrate the same increase in an explicit focus on the self in philosophy. This trend is "noticeable in Kant, central in Fichte, obtrusive in Kierkegaard, and pervasive in twentieth-century existentialism and personalities."[15]

HOW THE DIVERSIFICATION OF ROLES INCREASES INTERIORIZATION

A basic premise of social scientific thought is that concomitant with the transition from premodernity to modernity—which entailed increased economic specialization and state bureaucratization—came a diversification of and expansion in the number of social roles. The sociological literature contains classic portrayals of this change. The best known is probably the distinction made by the French sociologist Émile Durkheim. He contrasted "mechanical solidarity" (role differential is minimal, and roles are more or less interchangeable) with "organic solidarity" (occupationally specialized, interdependent, emergence of the division of labor). Resonating with Durkheim's insights was the distinction the German social theorist Ferdinand Tönnies (1855–1936) made between *Gemeinschaft* (pre-industrial "community," based on cooperation and custom, emotionally integrated) and *Gesellschaft* (industrial "society," based on law and public opinion, fragmented and artificial). We should also mention the American anthropologist Robert Redfield's (1897–1958) distinction between folk societies (small, isolated, and homogeneous, fact-to-face relations) and urban societies (large, secular, heterogeneous, impersonal relations).

The premodernity-to-modernity transition followed a chain of developments: The more economic complexity, the more employment specialization; the more specialization, the more roles the socioeconomic order requires; the more role-taking, the more personas an individual presents; the more personas, the wider are the gaps between these different presentation of selves (i.e., "I's" and "me's"), thereby leading to more reflexivity. Role-taking is perspective-making, and perspective-making increases the inner vistas of the psyche. All these changes were driven by utopian aspirations for a better future:

> A society whose entire political system, and indeed whose cosmology and moral order, is based in the last analysis on economic growth, on the universal incremental Danegeld [the coerced "tribute" paid by the English to the Vikings] and the hope of a perpetual augmentation of satisfactions, whose legitimacy hinges on its capacity to sustain and satisfy this expectation, is thereby committed to the need for innovation and hence to a changing occupational structure.[16]

Actually, the premodernity-to-modernity transition is not the only instance of increased economic diversification leading to a growing number of social roles, which in turn leads to deepening self-reflexivity and expanding interiority. The linkages among these phenomena are evident anytime the production of wealth transforms a person–property regime. For example, during the late Middle Ages in Europe, socioeconomic changes created a menu of choices that was previously unimaginable. Individuals could now choose between various occupations—knight, scholar, administrator, or monk. And each occupation lacked a clear script. The awareness of conflicting values burdened the individual with personal decision-making as well as questioning the sociopolitical established order. People had to balance their own self-interests with those of others while being true to themselves.[17]

Let us recap the main points in this section. As a general rule, the more social roles (or "me's") we must perform due to the pressures of sociopolitical and economic complexity, the weaker hierarchical relations become (though the *number* of social hierarchies has increased in modern times). This results in increased self-authorization as decision-making becomes more centered "within" the individual rather than in collectivities, institutions, or others. Increasing interiority, then, is a function of the multiplicity of parts we play in the numerous hierarchies that characterize modern times. These processes can be formulaically represented, where "*n*" equals the number of "me's" or social roles, and "I" stands for the imagined analog of the individual (whose value is always "1"):

Social roles = ("me's" x *n*) ÷ ("I" x 1)

Simply stated, as social complexity increases, the more roles are produced. This production of parts demanded by society in turn amplifies conscious interiority.

NOTES

1. Rosenthal, 1984, p. 21.
2. Nørretranders utilizes the idiom of "I" and "me," but his purpose is somewhat different from my own: "The role of the *I* in learning is precisely to force the nonconscious, the *Me*, to practice, rehearse, or just attend. The *I* is a kind of boss who tells the *Me* what it must practice. The *I* is the *Me*'s secretary" (Nørretranders, 1998, p. 303).
3. Mead, 1967, pp. 654–56, 609–10.
4. Jaynes, 1976, p. 63.
5. Goffman, 1959, pp. 80–81.
6. For further discussion of the relation between the "I" and "me," see "The Self as Interiorized Social Relations: A Jaynesian Approach to Problems of Agency and Volition" (McVeigh, 2006a).
7. Acting, incidentally, requires more than an audience; it also requires an interiorized self. Also note that the "same theatricalizing genius is at work in us when we recall the events of our past (no matter whether it is the events of yesterday or of 10 years ago). We cause these events to unfold again before our mental eyes. In other words, we compose a historical play, nondramatic in character. We stage it, and we ourselves appear in it as spectator and critic" (Evreinoff, 1990, p. 423).

8. Role proliferation, resulting from social complexity, is the primary historical driving force behind "I–me" mirroring. However, more immediately in terms of childhood socialization, self-reflexivity emerges from scrutinizing the actions of others: A "greater understanding and more acute observation of others may permit by comparison deeper understanding of the subjective self" (Benton, 1982, p. 269).

9. Barfield, 1967, pp. 171, 169.

10. Contrary to what researchers who are so impressed with chimps looking at themselves in mirrors might believe.

11. Rosenthal, 1984, pp. 8, 11, 12.

12. Rosenthal, 1984, p. 14.

13. Campbell, 1987, p. 73.

14. Campbell, 1987, p. 71.

15. Ong, 1982, p. 178.

16. Gellner, 1983, p. 32.

17. Morris, 1972, pp. 121–22.

Chapter Nineteen

The Birth of Modern Psychology

THE NINETEENTH-CENTURY'S DISCOVERY OF THE MIND

We take it for granted that thought itself can be the stuff of experimental science.[1] But this assumption is actually part of an intellectual revolution that is still being played out as we enter the twenty-first century: a progressively sharper body–mind dualism that segregates the psyche from the physical, observable world. In an attempt to come to terms with these intellectual developments, new lines of inquiry have been pursued. At the same time attempts were made to investigate the soul from a more scientific perspective. This reinforced interiorization. Eventually, psychological spatiality, or mind, became a focus of study in and of itself, leading to a new discipline by the late 1800s that centered on the introspecting individual: Psychology. In many ways the "incorrigibly private and individualized" person was itself the "product of a highly successful late nineteenth-century scientific research program in sensory physiology and experimental [P]sychology." In this chapter I highlight the key developments that delineate the birth of modern research Psychology from the angle of increasing conscious interiority.

WILHELM MAX WUNDT: THE FATHER OF PSYCHOLOGY

Wilhelm Max Wundt (1832–1920) is called by many the "first Psychologist." Before Wundt physiologists and philosophers certainly had an interest in Psychological issues, but they were not "Psychologists." With Wundt this discipline came into its own, freed from the speculation and spiritualism of religion, the materialism and animistic energies of medico-physiology, and the skepticism, abstraction, and meta-framing maneuvers of philosophy. He is credited with, among other things, delineating the first scientific experimental Psychology (thus the appellation "father of experimental Psychology"). Mental processes would be studied as mental processes, not as reflections within the soul, not as bodily sensations, and not as abstruse speculation.

Wundt studied under Johannes Müller and Emil du Bois-Reymond at the University of Berlin. Eventually he was appointed a lecturer in physiology at Heidelberg, where he worked for Hermann Helmholtz as a laboratory assistant and pursued research in physiological Psychology. In 1875, he was appointed professor of physiology at the University of Leipzig, where he was granted a room to store his research equipment. Four years later he conducted his first experiment in his Institute for Experimental Psychology, which is considered the first genuine laboratory of modern scientific Psychology where research on sensation, perception, vision, auditory, touch, and time-sense was carried out. This became a model for other laboratories. Wundt's most important ideas were expressed in his *Grundzüge der Physiologische Psychologie* (1873–74), which was translated into English by his student Edward Titchener (1867–1927) as *Principles of Physiological Psychology* (1904).

Wundt argued that Psychology was not reducible to physiology or any other discipline. It is the "science of experience" (*Erfahrungswissenschaft*) and should be divorced from metaphysics. However, for Wundt, "experimental" was practically synonymous with "physiological" since he believed that, though Psychology deserved its independence, it was still very much informed by physiology. Wundt was neither a reductionist nor a dualist. The goal of Psychology should be to bridge the dualism between the interior (feelings and the consciously subjective) and the exterior (perceptions that refer to objects). An inner sense does not exist, and the distinction between "inner" and "outer" is ultimately not valid. More specifically, the problems of Psychology should include: (1) the analysis of conscious processes into elements; (2) the determination of the manner of connection among these elements; and (3) the determination of the laws of connection. He intended to establish an introspectivist Psychology of what can be perceived, i.e., the "contents" (*Inhalte*) of conscious interiority. He distinguished between "the field of consciousness" (*Blickfeld*) and "the focus of consciousness" (*Blickpunkt*). For Wundt, the subject matter and methodology are the same: introspective immediate experience whose data are *anschaulich* ("phenomenal"), although this German term "actually goes further in indicating the palpable nature of the stuff of experience." The defining methodology of Psychology is "introspection" or "self-observation" (*Selbstbeobachtung*), "without meaning anything about the self, or introspection without meaning anything about a mental eye that turns about and looks into the mind."[2]

On balance, the contributions of Wundt are at once strangely liberating and restrictive. He pushed a surprisingly narrow view of what constituted

Psychology—"the elements of psychic life"—and employed a rather strict and circumscribed definition of his methodology of introspection in his laboratory. If a subject could not report on introspective experiences, then Psychological experimentation could not be pursued. This is why he dismissed work in animal Psychology, child Psychology, or anomalous behavior, such as hypnotism or mediumship. And he attacked the more subtle and meandering style found in the very influential *The Principles of Psychology* (1890) by William James (America's "father of Psychology"), who searched for insights into the human psyche in religion, literature, and history, as well as in everyday experiences. And yet later in his career, Wundt would show himself quite open to the rich varieties of human experience, as evident in his grandly ambitious ten-volume project *Völkerpsychologie: eine Untersuchung der Entwicklungsgesetze von Sprache, Mythus und Sitte* (*Ethnocultural Psychology: An Investigation of the Developmental Laws of Language, Myth, and Customs,* 1900–20). He investigated a stunning array of ethnographic and social Psychological materials, believing that certain aspects of the human condition, such as "higher mental processes," could not be studied through experimental methods. Like the social sciences in general, Wundt in the end was torn between viewing Psychology as a *Naturwissenschaft* (natural science) or a *Geisteswissenschaft* (science of the human spirit in all its complexity).[3]

THE WÜRZBURG SCHOOL AND "IMAGELESS THOUGHT"

Eventually some of Wundt's opinions on the mind would be challenged. For example, a former student and Psychologist and philosopher, Oswald Külpe (1862–1917) questioned some of his mentor's key ideas.[4] Using trained individuals who would introspectively self-observe, Külpe and his colleagues found that though subjects would report an introspective experience, it was void of imagery. To phrase it metaphorically, their interiorized space was empty of any mental furniture. At the time, this caused somewhat of an intellectual earthquake since it was assumed that all thought should be accompanied by images of some sort—i.e., ideas were defined in imagistic terms. The phrase "imageless thought" was coined to delineate the subtleties of interiorized experience. What other type of "content" could there be in the mind if mental imagery were absent? The Würzburg School, as Külpe and his disciples became known, used the term "conscious attitude" (*Bewusstseinslagen*) to describe obscure, indescribable, and unanalyzable contents, which were "neither sensations nor ideas."

Similar to conscious attitude is *Bewusstheit* or "awareness," a term used by Narziss Ach (1871–1946) to indicate a vague, not easily described, intangible and impalpable "something there" that lacks images or sensation.[5] In an attempt to accurately delineate the different stages of a psychological act, Ach discussed the need for "systematic experimental introspection," meaning that a careful scientific technique was required despite the fuzziness associated with interiorized experiences. Along with Henry J. Watt (1879–1925), Ach used the concepts of *Aufgabe* ("conscious task"; or an instruction given to a person) and "determining tendency" (or "mental set"), meaning the preparation—by nonconscious cognition—of the execution of a task. It became clear that the tasks given orchestrated the associations and skills into a purposeful orderly sequence of behaviors outside of awareness.

The students and colleagues of the Würzburg School followed up on the implications of imageless thought. In deceptively simple experiments, Karl Marbe (1869–1953) would ask subjects to lift two weights and then judge which was the heavier. He found that, though a sense of introspection transpired and they assessed correctly which was heavier, an actual "judgment" was not mentally "seen" by the subjects. Their assessments spontaneously appeared, though they could not describe the "conscious content" that accompanied such assessments. Judgment, considered the defining trait of the purposeful, rational, and conscious individual, occurred while the subjects were introspectively blank. Here we might also mention Karl Bühler (1869–1963), who researched the varieties of subjective experience, such as doubt, surprise, and the "consciousness of consciousness," i.e., the meta-framed feelings of interiority.

HIGHLIGHTING INTROSPECTION AND INTENTIONALITY

That interiorization was emerging as a new sociopsychological phenomenon is evident in the work of Franz Brentano (1838–1917). A philosopher who wrote on politics and ethics, Brentano took a principled and courageous stand against the Church by disagreeing with the dogma of papal infallibility and resigned from the priesthood. For this, he had to step down from his professorship at the University of Vienna. His rejection of conventional religion caused an existential crisis that led him on a search for an ultimate reference. He found it in subjective experience, i.e., conscious interiority. His best known work is *Psychology from the Empirical Standpoint* (1874; republished in 1911 and revised as *On the Classification of Psychic Phenomena*).

Brentano's Psychology was not experimental but was grounded in empiricism. His empiricism meant that experience—or conscious interiority—is an unavoidable feature of the human condition, and it is from here that an epistemology of Psychology must begin. By "experience" Brentano wanted to transcend the distinction between two different types of experience separated by dualism: the introspected contents of mind and perceived phenomena that constitute our empirical experience of reality. He overcame this distinction between "inner" and "outer" experience by linking the subject and object of experience in the "act." Our minds, then, are fundamentally intentional in the sense that they demand direction or an object toward which they are oriented. Three types of acts can be discerned: (1) ideating (sensing, imaging); (2) judging (acknowledging, rejecting, perceiving, recalling); and (3) the psychic phenomena of loving and hating (wishing, resoling, intending, desiring).[6]

The legacy of Brentano is not as clearly delineated as the work of more experimentally oriented researchers who wrote at the same time. His writings have a subtlety that is often lost in the more cut-and-dry accounts of mind. Nevertheless, the "act school," stressing a top-down, purposeful, intentional, and individual-centered experiential perspective, had a lasting impact on Gestalt Psychology. The latter school, associated with Max Wertheimer (1880–1943), Kurt Koffka (1886–1941), and Wolfgang Köhler (1887–1967), focused on problems of perception that defy the more atomistic approaches of conventional psychophysical research. The act school also influenced the development on the ultrasubjectivity of phenomenology (a sort of interiorization of interiority) as seen in the work of Edmund Husserl (1859–1938). The enduring lesson of Brentano's work? Breaking psychological experience down into elemental components comes at a price, as it turns the mind into an aggregate of unconnected parts.

Alexander Bain

Though many contributors from the Anglophone world could be cited, I have selected Alexander Bain (1818–1903) for brief mention as he is often overlooked in histories of Psychology. Bain, who held a chair of logic at Aberdeen, was not a Psychologist though his contributions substantially developed the disciplinary autonomy of Psychology. His work spanned much of the nineteenth century, and he "stands exactly at a corner in the development of Psychology, with philosophical Psychology stretching out behind, and experimental physiological Psychology lying ahead in a new direction."[7] His *The Senses and the Intellect* (1855) and *The Emotions and the Will* (1859), repeatedly revised, remained a standard British Psychological text until the early twentieth century. In 1872 he wrote *Mind and Body* and in 1876 founded *Mind*, the first Psychological journal. He linked physiology to Psychology and investigated reflexes, instinct, and voluntary action. Though he never used "psychophysical parallelism," this was the premise from which he viewed the mind–body problem.

CONCLUSION: THE RECOGNITION OF CONSCIOUS INTERIORITY

In the same way that Psychology had come into its own as a recognized independent discipline, interiority had come into its own as an autonomous experience of the human condition by the late nineteenth century. Though researchers still confuse interiority with perception, thinking, reasoning, or anything associated with the especially fuzzy rubric of "consciousness," at least some thinkers have acknowledged it as *sui generis*.

NOTES

1. Datson and Galison, 2007, pp. 45, 263.
2. Boring, 1929, p. 327.
3. See also Wundt's *An Introduction to Psychology*, 1973 (1912); *Outlines of Psychology*, 1969 (1897); and *Elements of Folk-Psychology*, 1916.
4. Külpe wrote on aesthetics, ethics, and medicine, and founded the Würzburg School of experimental Psychology. Some of his better known works include *Grundriss Der Psychologie* (*Outlines of Psychology*, 1893), *Einleitung in Die Philosophie* (*Introduction to Philosophy*, 1895), and *Vorlesungen über Psychologie* (*Lectures on Psychology*, 1922).
5. Boring, 1929, p. 396.
6. Boring, 1929, p. 351.
7. Boring, 1929, p. 231.

Epilogue

Visualizing New Vistas of Modernity and Selfhood

Chapter Twenty

The Therapeutic Turn

EXPLAINING A MENTAL TRANSFORMATION

In this chapter I explore the birth of the "therapeutic society" by reviewing some recent works that take up this issue: Eva Illouz's *Saving the Modern Soul: Therapy, Emotions, and the Culture of Self-Help* (2008); Eva Moskowitz's *In Therapy We Trust: America's Obsession with Self-Fulfillment* (2001); and Micki McGee's *Self-Help, Inc.* (2006). The idea of a thoroughly therapeutized society, as presented by Illouz, Moskowitz, and McGee, is not new. Well-known theorists have interpreted the rise of the therapeutic worldview—e.g., Philip Rieff's *The Triumph of the Therapeutic*, Lionel Trilling's *Sincerity and Authenticity,* Martin L. Gross's *The Psychological Society: A Critical Analysis of Psychiatry, Psychotherapy, Psychoanalysis and the Psychological Revolution.*

The three aforementioned works all have merit, but they strike me as tautological. Thus, we need a more comprehensive theoretical perspective that takes fundamental changes in psyche into account. Along with Jaynes, I contend that increasing conscious interiority, signaling deeper historical changes, explains these social changes. My commentary on several books continues the exploration of the psychologized society. My intention is to illustrate how the arguments in these books resonate with Jaynes's definition of conscious interiority, particularly as it has expanded in the last century or so (chapter 21 picks up some of these themes).[1] For Jaynesian Psychology, historical contextualization is crucial, because conscious interiority, after it had entered the historical stage in the late second millennium BCE, underwent developments depending on specific periods and local conditions—i.e., depending on a society's political economic needs, certain features of conscious interiority were emphasized or de-emphasized.

A CHANGE OF MIND IN THE NINETEENTH CENTURY: SOME HISTORICAL PERSPECTIVE

A salient aspect of the psychologization of society is what Eva Moskowitz calls the "therapeutic turn," which, according to her, began in the 1850s. Moskowitz chronicles how faith in an inner happiness has been intimately bound up with major sociohistorical developments; for her it is clear that a transition occurred from a less-therapeutic mind-set to the highly psychologized society of today.

Moskowitz begins her tale of the birth of the psychological society by introducing Phineas Pankhurst Quimby (1802–66) in the first chapter of her book ("Illness: A New Cure, A New Faith, 1850–1900"). In 1859 Quimby began to advocate "Spiritual Science" that, for the time, was based on novel notions, e.g., "investigations in [P]sychology," "mental therapeutics," and "mind cure." It may be difficult for us to appreciate, but before 1850, medicine was interpreted using either a somatic or spiritual lens, so that treatment was not conceived as something explicitly psychological.[2] For Quimby, however, one gained power over disease by understanding the mind. Conventional medicine neglected the role of the psyche while religion contributed to sickness. Prayer was secularized, and the pursuit of personal happiness became a commandment. Not sin, but pessimism and negative thoughts were to be avoided. Religion and science were fused in what develops into what we call the Psychological.

Quimby's ideas constituted an important precursor to the postmodern therapeutic ethos of "mandatory happiness." In particular, a movement of "religious medicine" loosely based on his teachings, called New Thought, would emerge.[3] Mental suggestion, psycho-gymnastics, and "mental photography" were aspects of this new worldview. The latter, developed by Henry Wood in his *Ideal Suggestions through Mental Photography* (1893), attempted to make our introspectable contents publicly visible to the physical eye (rather than just the individual's private "mind's eye"). Pictures of a placard would be taken on which one would write down positive thoughts, and then an individual would stare at it for ten to twenty minutes. This practice was a "key aspect of the modern therapeutic agenda" in which one's mental processes and deep-seated emotions were made visible.[4] Organizations such as the League for the Larger Life, National New Thought Alliance, and the International Divine Science Association were founded, and eventually over one hundred New Thought magazines would sprout up.

For her part, Micki McGee contends that an "inward turn . . . has been underway since the early eighteenth century." This "move toward a greater sense of interiority and a focus on the self has been a tendency of modernity." In her first chapter entitled "From Calling to Vision: Spiritual, Secular, and Gendered Notions," she explores the ethico-religious grounding of self-help by noting that the premodern idea of a "calling" was a "dual concept, split between spiritual, otherworldly dimensions and the material, workaday world."[5] Later developments witnessed a shift from calling to "self-actualization." The latter was a Romantic notion that regarded one's life as a creative, aesthetic enterprise. Expressive individualism, something "joyful and playful," became self-discovery. Eventually self-actualization became associated with a "therapeutic turn" and a highly psychologized view of the person. As McGee phrases it, spiritual values have been recast as "therapeutic theism."

THE "THERAPY-CENTRIC SOCIETY"

Moskowitz lists three tenets of the "therapeutic gospel": (1) happiness should be our supreme goal. Wealth, public recognition, and high moral character are only valuable if they make us happy; (2) our problems stem from psychological causes; and (3) psychological problems underlying our unhappiness are treatable.[6] "No other nation [U.S.] in the world puts so much faith in emotional well-being and self-help techniques." She suggests optimistic individualism, generated by prosperity, accounts for such dedicated "soul-doctoring." This very optimism, however, "seems continually undermined by consumerism," which "markets psychological ill-being." Products, and even the "act of purchasing itself are the cure." Other reasons for this "national investment in feelings" include the: (1) decline of religion; our loss of faith in salvation has been replaced by self-glorification; (2) institutional experimentation that has led to the growth of psychologized welfarism; (3) the "dynamic nature of America's cultural economy"; and (4) the demands of corporate culture, which include not just production, but reproduction, i.e., how to bring up one's children as workers while avoiding dysfunctional families.[7]

Eva Illouz also investigates how a "therapeutic discourse" constitutes a "formidably powerful and quintessentially modern way to institutionalize the self."[8] Significantly, Illouz stresses how we should be sensitive to the context in which therapeutic discourses occur, e.g., corporation, marriage, support groups. And she rightfully criticizes Michel Foucault and like-minded thinkers who do not take the "critical capacities of actors seriously." The self, as the "prime site of the management of the contradictions of modernity," requires techniques—modern psychotherapy—to manage those contradictions. "In other words, [P]sychology is less about Foucaultian 'surveillance' or 'bio-power' than it is about containing and managing the contradictions of modern selfhood." Illouz makes another important point: A basic tenet of the therapeutic persuasion is that "what one feels" needs to be revealed, dissected, and discussed. The role of a therapist is to expose, correctly name, and transform emotional content of conscious interiority. However, because affective experience is fluid and contextually generated, there is no emotional substance necessarily waiting to be revealed. Moreover, a "competent emotional response does not necessarily entail a self-conscious awareness of one's emotional responses."[9]

In her incisive survey of American self-help literature titled *Self-Help, Inc.*, McGee historically contextualizes the meaning of this "inward turn," showing persistent continuities as well as divergences. She skillfully disentangles the complex ideological tapestry of religious traditions, psychological ideas, and medico-therapeutic approaches that, woven together, comprise our current understandings of the self. She is particularly concerned with how the rapidly changing political economic climate is cultivating this influential genre. McGee explores how the allure of such literature is more recently tied to economic insecurity, though in a hopeful manner she writes that such an examination "might be mined for political opportunities."[10]

THE THERAPEUTIC THEISM OF THE TWENTIETH CENTURY

Moskowitz's book treats a number of social realms which were impacted by therapeutic theism, illustrating just how widespread this complex of ideas has become. The first social realm, detailed in the second chapter of her book ("Poverty: Reformers Offer Treatment, 1890–1930"), deals with the intersection of social problems (e.g., poverty), the state, and social reformers who, imbued with the therapeutic gospel, encouraged the psychologization of welfare. Some reformers traced the roots of poverty to character flaws, and saw asthenontology—the study of human weakness—as the answer. Others sought to banish punishment from the criminal-justice system by injecting a "clinical atmosphere into the courtroom"[11] and by replacing sentencing with diagnosis.

Another area of social life impacted by therapeutic theism was marriage. According to Moskowitz, in the 1920s and 1930s, marriage began to be judged by a new psychological valuation that placed a premium on happiness. An increasing divorce rate caught the attention of commentators. Ernest R. Groves authored *Marriage* (1933), the first textbook on matrimonial relations, and along with Phyllis Blanchard, wrote the first book on social psychiatry (*Introduction to Mental Hygiene*, 1930). Other works, such as L. M. Terman's *Psychological Factors in Marital Happiness* (1938), appeared, as did new journals, with names like *Marriage Hygiene*. In 1942 the American Association of Marriage Counselors was established.

Military Psychology emerged due to the large number of psychiatric casualties caused by the industrial-scaled, mechanized massive killings of twentieth-century warfare. Martial issues constitute another social realm configured by psychologization.[12] While a limited program was provided in World

War I, during World War II millions of American men for the first time were exposed to the tenets of the psychological gospel. Theodore Adorno et al. attempted to understand the mental underpinnings of fascism in their *The Authoritarian Personality* (1950), and eventually, the need for "psychological warfare" in order to predict and understand the psyche of our enemies would be recognized.

After the war the American home became the next social realm targeted for the proselytizing of the therapeutic gospel. Moskowitz explains how attempts, driven by the consumption engine of postwar prosperity, were made to push products by appealing to women's apprehensions and aspirations.[13] The virtues of self-fulfillment, linked to domesticity, were put on display in women's magazines, which introduced a new idiom—e.g., unconscious, psychosomatic, defensive reaction. Other terms that we now take for granted—ego, inferiority complex, self-esteem—became household words. Meanwhile, concerned with the "happiness" of the average American, the federal government stepped in with the 1946 National Mental Health and 1955 National Mental Health Study Acts. In 1949, the National Institute of Mental Health was established, as was the Joint Commission on Mental Illness and Health in 1955.

Finally, some other social realms impacted by psychologization include the counterculture movements, feminism, and "personal politics" can be understood within the context of the dramatic growth of Psychology as a discipline after 1945. The humanistic Psychology of Carl Rogers and Abraham Maslow would play an important role in these developments by stressing the "liberation of the self."[14] This was also the period when the psychological harm of racism was recognized—e.g., Abram Kardiner and Lionel Ovesey's *The Mark of Oppression: A Psychological Study of the American Negro* (1951). How modern society idolizes the individual's inner life seemed to have become the point of existence. Encounter meetings, support groups, and self-help books of the "me" generation advise us to articulate a more authentic self and display our true feelings.[15] Personal problems are made public via technologies such as "TV talking cures" and the Internet that were inconceivable to earlier generations. Identity politics, a product of the 1960s and 1970s, became linked to the therapeutic gospel, as did the identity of being an addict (now we can be addicted to anything: sex, love, substances, shopping). Addiction has spawned other types of self-identity, e.g., codependents and survivors. By the 1980s a change in therapeutic thinking was discernible. "Rather than focusing on human potential, promoters of the therapeutic gospel focused on psychological vulnerabilities."[16]

CONCLUSION: "A RELIGION OF SELFHOOD"[17]

Rather than God, Americans now turn to "psychological idolatry."[18] The therapeutic practices Illouz investigates are charged with a long list of tasks: to build coherent selves, procure intimacy, provide a feeling of competence in the work world, and facilitate social relations. This discourse has thoroughly penetrated every nook and cranny of daily life and possesses a strong institutional basis in university departments, research institutes, journals, media, and a massive army of Psychologists and counselors. Consequently, the therapeutic ethos has come to "constitute one of the major codes with which to express shape, and guide self-hood."[19]

NOTES

1. This chapter borrows from reviews of these books I wrote for Metapsychology Online: McVeigh (2006b, 2008, 2009).
2. Moskowitz, 2001, p. 280.
3. Moskowitz, 2001, p. 19.
4. Moskowitz, 2001, p. 23.
5. McGee, 2005, pp. 47–48, 76–77, 26.
6. Moskowitz, 2001, p. 3.
7. Moskowitz, 2001, pp. 278, 279, 280.
8. Illouz, 2008, p. 6.
9. Illouz, 2008, pp. 4, 243, 206.
10. McGee, 2005, p. 12.
11. Moskowitz, 2001, p. 45.
12. Moskowitz, 2001, chapter 4, "War: The Soldiers Psyche, 1941–1945."
13. Moskowitz, 2001, chapter 5, "Home: The Unhappy Housewife, 1945–1965."
14. Moskowitz, 2001, chapter 6, "Social Protest: Liberating the Psyche, 1960–1975."
15. Moskowitz, 2001, chapters 7 and 8, "Feelings: Expressing the Self, 1970–1980" and "Personal Problems and Public Debate," respectively.
16. Moskowitz, 2001, p. 282.
17. The name of a sermon written by Julian C. Jaynes, Jaynes's father. The latter read this sermon in 1985 when he was invited to speak at the one hundredth anniversary of his father's ordination (Woodward and Tower, 2007, p. 52).
18. Moskowitz, 2001, pp. 2–3.
19. Illouz, 2008, p. 6. Notably, the "new middle class," especially among women in this group, is particularly receptive to this ethos.

Chapter Twenty-One

Self-Idolatry

The Dark Side of the Psychotherapeutic Society

Have many times have we heard a relative complain that their children's schools have enacted Alice-in-Wonderland policies designed to enhance "self-esteem": Teachers are not allowed to correct with red pens because of the negative associations of their color; sports teams are not allowed to win because the defeated experience disappointment; everyone is awarded for excellence so no one feels dejected; and students are encouraged to have an inflated sense of their academic abilities. How widespread such practices are is debatable, though certainly the perception exists that "feelings are sacred and salvation lies in self-esteem." In other words, the contents of the self deserve special care and consideration. This psychologization of society is also evident in the dramatic increase in the number and categorization of mental health maladies. Consider that in the fifteen years that the *Diagnostic and Statistical Manual of Mental Disorders* has been available, the number of mental disorders listed has grown from one hundred to more than three hundred. Between 1945 and 1970, the number of psychiatrists increased from 3,600 to 18,400, and between 1957 and 1976, those seeking Psychological help increased from 14 percent to 24 percent.[1]

Our lives are more and more supervised by psychiatrists, psychoanalysts, clinical Psychologists, psychotherapists, social workers, psychiatric nurses, guidance counselors, marriage and family therapists, educational Psychologists, lay therapists of all stripes, and the assorted self-help gurus, maharishis, and spiritual leaders. Encounter groups, workshops, and pastoral counseling round out this army of psycho-engineers intent on analyzing, evaluating, and calming the storms and stresses of conscious interiority. The ranks of those engaged in psycho-engineering encourage us to believe that we should be "dependent on others for guidance as to what is real or false." According to one observer, the mental health industry is creating its own patients.[2] The psychologization of society is also apparent in how the last quarter of the twentieth century has witnessed a dramatic rise in sales of "self-help" books: between 1972 and 2000, the number of self-improvement books more than doubled. Moreover, this genre constitutes an influential portion of the publishing industry, and one-third to one-half of Americans bought such works.[3] Martin L. Gross, in his "Critical Analysis of Psychiatry, Psychotherapy, Psychoanalysis and the Psychological Revolution," noted that it is not just the "external transformations" of the human condition, but the "internal shift" in our psyche that has "altered both our actions and expectations more than any technological force."[4] Never before has a civilization been so preoccupied with the self.[5] This ideological tidal wave has impacted art, science, religion, philosophy, our moral codes, and lifestyle.

How did we reach an "age consumed by worship of the psyche"?[6] What explains this self-idolatry? What does this say about modern societies? Our relations to our political economic institutions? Our relations to ourselves?

In this chapter I explore the "dark side" of the psychologized society. My premise is that interiority underwent major changes during the nineteenth century whose impact we are now experiencing (e.g., "therapeutic turn"). This theme can be subsumed under an overarching argument: history is witnessing an ever-increasing conscious interiorization of psyche.

THE PRICE OF THE PSYCHOLOGIZED SOCIETY

The idea that we increasingly live in a psychologized, self-centric world has been a target of commentary for quite some time. Philip Rieff's important *The Triumph of the Therapeutic: Uses of Faith after Freud* 1987 [1966] explored the easy, selective, and relativistic ethics that characterize our historical period in which the certainties of religious truths have been replaced by the feel-good, mushy morality of psychologism. In *The Rise of The Therapeutic State* (1991) Andrew J. Polsky contended that "marginal" citizens are increasingly managed by government agencies who attempt to "normalize" their attitudes and behaviors. James L. Nolan, in *The Therapeutic State: Justifying Government at Century's End* (1998), argued that the institutions of education, criminal justice, welfare, political rhetoric, and civil law have acquired a therapeutic agenda. In *The Therapeutic Corporation* (1999), James Tucker noted how therapy has become the most common way of handling conflict in the postbureaucratic work environment. Steven C. Ward's *Modernizing the Mind. Psychological Knowledge and the Remaking of Society* (2002) has taken up similar themes, and in *Therapy Culture: Cultivating Vulnerability in an Uncertain Age* (2004), Frank Furedi has persuasively challenged the usefulness of the "therapeutic turn." He believes that every facet of daily life has fallen under the spell of psychotherapy, resulting in a new conformity and unhealthy management of emotions. Consequently, we are socialized to have feelings of powerless and alienation.

Alienation characterizes modernity. Unlike premodern versions of alienation, which meant being separated from God or one's immediate community, its secularized, psychologized, modern sense means being disaffected from one's self. Arguably, much of present-day disaffection can be traced to "mandatory happiness," and despite the positive, optimistic, and hopeful language of the therapeutic worldview and the self-help literature, a less cheerful side to the psychologization of society should be highlighted. When people are encouraged to "feel about how they feel," we know there is undue attention being directed at the self. Moskowitz points out how the "gospel of psychological happiness" distorts our appreciation of the production of self-disaffection by "underlying economic and political realities," the "realities of class," and "flaws in the market economy." Rather than "emotional retooling," "remedial emotional education," or "emotional illiteracy," we should acknowledge the root social conditions that generate alienation. For example, the discourse on what makes for a happy marriage in fact represented "deep-seated gender and class interests," and despite well-intentioned efforts of reformers, such attempts might end up ignoring larger economic conditions and "psychologized difficulties that were structural and social."[7]

The late nineteenth century witnessed the emergence of a discourse of personal disaffection, but it was not until the twentieth and twenty-first centuries that its treatment became institutionalized in counseling, religious centers, clinics, and the medical establishment. Alienation has many causes and manifestations: narcissism, constructing a false self; collapse of moral hierarchies; intense privatization of life caused by capitalist anomic organization; the emotional emptiness of being severed from communal relationships; surveillance and state control; and atomistic individualism. Terms such as "therapeutic self-absorption," "therapeutic ethos," the older "talking cure," as well as practices such as commercial self-help books, support groups, assertiveness training programs, and television programs providing "one-show" counseling testify to attempts to treat alienation.

Alienation often results when we are asked to perform contradictory roles—e.g., we should be self-reliant, yet attuned to the needs of others; "conduct relationships in a highly rational way, yet be highly focused on our own and others' emotions; be a unique individual, yet constantly cooperate with others." The therapeutic persuasion advocates a disengaged self that should be busy with self-mastery and control as well as a sociable self that brackets "emotions for the sake of entering into relations with others." And while the individual "dwells excessively on his or her emotions, he or she is simultaneously required not to be moved by them."[8] To a large degree such incongruities can be traced to tensions caused by emotional control for the sake of organizational efficiency and the "commodification of emotion" (see chapter 3 in Illouz's book: "From *Homo economicus* to *Homo communicans*").

As counterintuitive as it is for many of us, more intimacy leads to alienation. This is the point of a chapter entitled "The Tyranny of Intimacy" in the book by Illouz introduced in an earlier chapter. The "ideology of pure emotion," and its associated values of openness, authenticity, and informality, becomes a social burden (and as someone who lived for sixteen years in an Asian society that does not see the benefits of uninhibited directness, I have concluded that let-it-all-hang-out casualness is an overrated virtue). For Victorians, intimacy involved finding the appropriate person to whom to reveal one's "true self"; revelation itself was not a problem. However, with so much confusion about what constitutes "authenticity," the act of uncovering one's self has become an intrusive burden for some and an unwelcome challenge by many.

We might mention another type of alienation that is the consequence of what Illouz calls "Triumphant Suffering." She discusses the psychodynamics of victimization and how therapeutic persuasion is based on the premise that moral problems are diseases to be cured. Consequently, a therapeutic culture has emerged whose mission is to heal the sting of "thwarted opportunities for self-development." Suffering and victimhood come to define the self, and "performing the self through therapy" turns private experiences into a narrative to be publicly consumed. Such a development, arguably, has demeaned and cheapened personal life. Therapeutic narratives, then, should be criticized for "making *too much* sense of one's life."[9] When we trade in our feelings of victimization to engage in gratuitous sentimentalization for the sake of sheer self-vindication and self-gratification, others become mere mirrors for our daily dramatizations. The purpose of genuine feeling is forgotten. Alienation results.

Significantly, the stuff of our inner selves has become so central in the present day that unhappiness (something psychological) has metaphorically mutated into a physical ailment (figure 21.1). Everyday worries are reconceptualized into a medical problem: Daily worries (metaphrand) are described as something that requires curative treatment (metaphier). The latter implies physical healing (paraphier) which generates the paraphrand of illness.

We equate happiness with normality, thereby assuming that any type of suffering—disappointment, routine frustrations, and other everyday emotional frictions—requires psychotherapy. "Natural emotions such as outrage, despair, grief, jealousy, suspicion, disappointment and passing depression are made to appear not only undesirable but abnormal." Normality has been redefined. The nasty, unhappy, and eccentric have always been with us, but

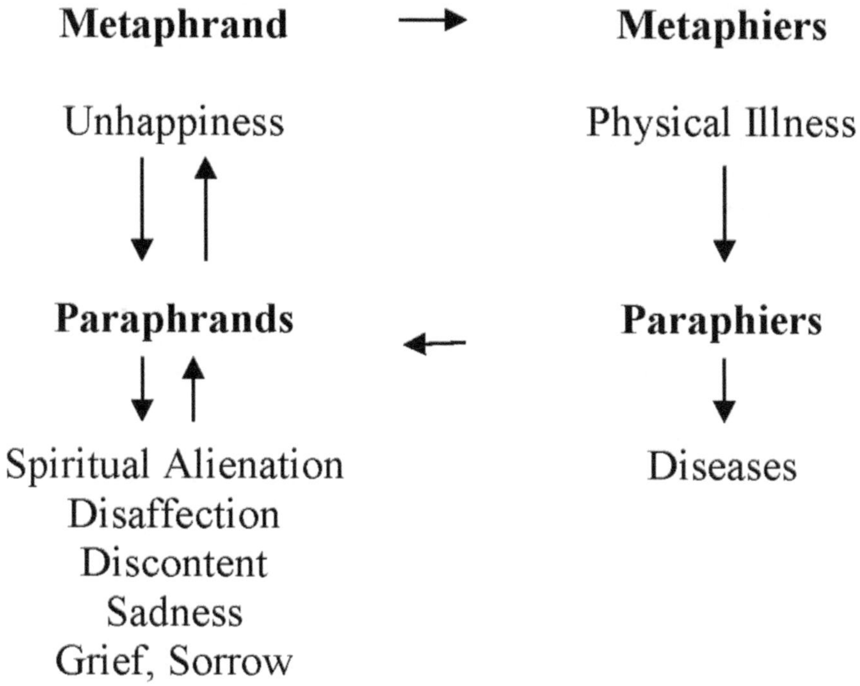

Figure 21.1. Alienation and Discontentment Reconceptualized as Sickness.

not the large number of those supposedly in need of Psychotherapy. The upshot is that the "boundary between the well and the sick has been blurred."[10] Moreover, immoral behavior and illness are often conflated.

THE LABORING AND BELABORED SELF

Related to the issue of work, of course, is class, and Illouz contends—demonstrating how psyche is very much a cultural construction—that one's socioeconomic standing configures one's inner world of feelings (chapter 6, "A New Emotional Stratification?"). Society's "emotional fault lines" are more salient between the working and middle class, rather than between men and women. The self in need of therapy, then, implicates property and wealth. Indeed, McGee comments on how the permeation of the market into every corner of social life results in the fusion of Psychological with financial counseling (think *The Suze Orman Show, Til Debt Do Us Part*, and other similar programs).[11]

McGee discusses how the very notion of work has acquired a new meaning: labor on and for the self. This has been a value and aspiration reserved for well-propertied males (we might also insert "white"). The "most striking features of the 'unisex' literature of self-improvement is the poverty of the solutions offered to women in their quests for self-made success." Chapter 4, "The Self at Work: From Job-Hunters to Artist-Entrepreneurs," elaborates on how a psychotherapeutic ethos permeates our society; how self-realization through work has come to signify personal salvation; and the "idea of the artist as an example for the postindustrial worker."[12] However, so much work on the self, whether spiritually or aesthetically understood, constitutes uncompensated labor.

McGee takes a critical stance vis-à-vis self-improvement literature, noting that the "promise of self-help" can enslave workers who become stuck on a treadmill—the self is not improved but ceaselessly belabored. Such a beleaguered self, which results in alienation, is "perennially at work on itself." Exploring the "cultural preoccupation with the self in terms of labor,"[13] McGee describes how an incessant "labor on the self" in the form of "immaterial labor"—i.e., mental and emotional work—whether on the job or for one's self-improvement, can be just as alienating. The troubling aspect of self-help literature is that if one's success results from one's own efforts, then the responsibility for failure is also one's own. Moreover, she contends that by overly focusing on the self, the individual is cut off from supportive collectivities, leading to an "increasingly isolated individualism." In "At Work on the Self: The Making of the Belabored Self" (chapter 5), she expands her book's themes, noting, for example, that the "self-authoring self," as ideally active, autonomous, and independent, should not passively rely on others. Such a sentiment, however, may lead us to ignore the contribution, support, and sacrifices of others.

It is a truism, as Illouz points out, that modern economics has encouraged two complementary processes: the "emotionalization of economic conduct" and the "rationalization of intimate relations."[14] So we should ask: Do we want to live in a society in which the values of the marketplace have been transplanted to the personal world of intimate life? While earlier ideas about work viewed it as service to the community and a source of sustenance, it has now become a matter of self-realization. Within this context we can question the meaning of work: What is its purpose? Self-expression? Personal fulfillment? A source of identity? In any case, work itself has become the sole source of self-fulfillment for many.

THE POLITICAL IMPLICATIONS OF A SELF-CENTRIC CULTURE

A final chapter in McGee's book, "All You Can Be, or Some Conclusions," rounds out her treatment, and here it is worth addressing McGee's sociopolitical commentary. Self-help culture, with its imperative to "be all one can be," is double-edged. The belabored self is "caught in a cycle of seeking individual solutions to problems that are social, economic, and political in origin." However, McGee optimistically opines that self-help culture may be used for "cultivating social change" and that it "might be tapped for a progressive, even a radical, agenda." So far, so good; these are not unreasonable suggestions. But McGee, who sets up a dichotomy between "capitalist" and "democratic" demands (predictability the former is a term of censure), runs aground when she writes that a "social safety net with a guaranteed minimum living allowance would be a necessary component of a politics of self-realization."[15] Such thinking, inspired by economic reductionism, leads to an ideological default mode premised on a naïve notion that social ills can be explained by material inequities; therefore, redistributive programs will fix society.[16]

Certainly the vagaries of postmodern capitalism demand safety-net and stepping-stone welfare. However, overgenerous welfarism fails not because it inadequately redistributes resources, but because it encourages its recipients to see little reason to "invest their selves" into property for which they did not work. Self-ownership, a species of self-realization, comes from self-directed accumulation of material security, not handouts. Social stability arises not from dividing the pie up more equally, but from encouraging individuals to self-justify their linkages to whatever size of the pie they obtain. The twentieth century, with its long list of failed socialist experiments, demonstrates the pitfalls of equalizing property relations (hard social engineering). Arguably, political economic regimes should foster an environment of opportunity by market regulation and dismantling obstacles to genuine self-ownership (soft social engineering). Moderate social engineering, as seen in social and liberal democratic welfare attempts, are less ambitious than socialist social engineering and their shortcomings not as glaring, but their results are decidedly mixed. In any case, economic measurements are objective, whereas how much wealth makes an individual content and how willing one is to pursue property accumulation are subjective and a matter of personal preference.

FINAL THOUGHTS: AN EMPIRE OF THE SELF AND THE "ESSENCE OF MODERN ACTORHOOD"

As Moskowitz, McGee, and Illouz make clear, self-idolatry, self-makeover culture, and therapeutic thinking are not just an American story; these are global phenomena (though arguably more evident in the United States because of its strong individualistic ethos and its vast wealth, which allows many of us the luxury and resources to pursue self-fixation and self-analysis). Illouz notes how an increasing number of societies are adopting the same economic and political models, as well as definitions of human nature, and Psychology, it appears, is one of the main pillars of globalization. Despite their local cultural inflections, different societies are using similar systems to "establish the essence of modern actorhood."

What is this "essence of modern actorhood"? Following Jaynes, I contend that a key element is increased conscious interiorization. But such a process can only be appreciated within the context of global consumerist capitalism. As economic changes continue to position the individual center stage—supersized egos, oversized waistlines, the dramatic rise in cosmetic surgery, a penchant for McMansions, tanklike SUVs, a self-indulgent lifestyle fueled by easy credit, deficits, debt, and "creative accounting"—a self-centric culture breeds narcissism. This is, incidentally, a theme investigated by Jean Twenge and W. Keith Campbell in their *The Narcissism Epidemic,* a work that continues well-established cultural criticism thesis (e.g., Christopher Lasch's 1979 *The Culture of Narcissism: American Life in an Age of Diminishing Expectations*). Meanwhile, technology relentlessly drives the expansion of the empire of the self, encouraging us to lay bare our private lives on Facebook, MySpace, Twitter, and YouTube at a head-spinning rate that some find disturbing. Gross, in a prophetic observation, noted in 1978 that "privacy has become a serious victim" in our "let-it-all-hang-out" culture. People

> are justly concerned about the invasion of privacy by data banks, credit checks, government agencies. Yet they are surprisingly willing to surrender their natural rights of privacy when a [P]sychological rationale is offered. They will tell their life stories to professionals, paraprofessionals, even laymen who will listen. They will spill out their secret resentment, problems and sex lives to a dozen strangers called a *group.* [17]

All these developments can be understood as "self-consumption": if both the self and consumption are central to our lives, the interior contents of the self—feelings, thoughts, preferences, and prejudices—become something to be savored, even at the expense of social relations.

NOTES

1. Moskowitz, 2001, pp. 2–3, 4, 217.
2. Gross, 1978, pp. 4, 17.
3. McGee, 2005.
4. Gross, 1978, p. 3. For Gross these changes have led to the "Psychological Idea" that denies three principles that have grave import for modern society: (1) the "organic view of the mind as a biochemical device with a relatively inflexible gene-determined inherited style"; (2) the role of religion (i.e., the self has replaced God); and (3) a behavioral model of morality, i.e., in previous times one judged oneself and others by what one did, not by what one thought (Gross, 1978, p. 13). Behavior was morally self-evident, so the need to search for complex motivations (e.g., shaped by the emotional pressure cooker that for many of us is the family) was largely absent.
5. Gross, 1978.
6. Moskowitz, 2001, p. 2.
7. Moskowitz, 2001, pp. 2–3, 7, 99, 69.
8. Illouz, 2008, pp. 243, 103, 104.
9. Illouz, 2008, p. 196 (emphasis in original).
10. Gross, 1978, pp. 17, 5.
11. McGee, 2005, p. 105.
12. McGee, 2005, pp. 79, 128.
13. McGee, 2005, pp. 12, 16.
14. Illouz, 2008, p. 239.
15. McGee, 2005, pp. 177, 24, 183.
16. It is worth noting that McGee offers political commentary on what might constitute an "effective politics" that would "necessarily embrace the demise of the public-private split . . . rather than mourning its passing" (p. 185). I wonder. Arguably in certain contexts, this proposition seems reasonable, but at the same time, it might let in a Trojan horse of the "public" (which, incidentally, must be clarified as designating either official state or unofficial civil society). With so many private sites now needing protection from public intervention, McGee's proposition must be entertained with extreme caution. Indeed, McGee asks "why the increase in the number of self-improvement groups and culture has far outpaced the growth of either progressive or radical political movements throughout the last part of the twentieth century" (2005, p. 189). Probably because many sense that, while self-help culture (private endeavors) and the political economic regime (public concerns) are certainly linked, they do not (and should not) necessarily align.
17. Gross, 1978, p. 13 (emphasis in original).

Chapter Twenty-Two

Modern Spatiality, the Soul, and the Psyche

Notions of spatiality develop unpredictably. While European thinking on geography was weighted down by sacred knowledge that would not be challenged in any practical way until the late 1400s, the religiously inspired interior worlds of visionaries and a type of emergent proto-Psychology began to be recorded in the twelfth and thirteenth centuries. New geographies were being discovered within each soul, and the individual was linked to God within the soul, rather than through the formal and visible layers of the clerisy. During the Middle Ages the alienation of "men's souls from the walled Church sent German mystics on an inward migration which was to continue until the Reformation." Mystics such as Meister Eckhart (1260–1328) often described their experiences in the form of a dialogue between the heart and the Godhead. Mechthild of Magdeburg (circa 1210–82), who between 1250 and 1265 penned her visions, takes the reader to Hell, Purgatory, and Paradise. These otherworldly realms are in fact described as introcosmic, inner kingdoms of the soul: "All the splendour of heaven and earth is to be found in the soul, the hiding place of God."[1] Alighiere Dante's (1265–1321) *Divine Comedy*, of course, is the most famous literary exploration of the intersection between the person's "inner kingdom" and the spiritual realm.

Eventually, celestial and terrestrial space became united into "one continuous physical domain." This new domain challenged the discontinuity of spiritual and physical realms, and space itself became regarded as monistic. The dualism that split corporeal existence from incorporeal presence was weakened. During the European Middle Ages "space was spiritually graded. In Euclidean space, however, one end of the universe is the same as the other, and Heaven and Hell become empty symbols."[2] Space as a "hierarchy of values was replaced by space as a system of magnitudes."[3] With the loss of the transcendent, only one kind of reality amenable to scientific investigation remained: it was physical, measurable, and mathematicalized. From the late seventeenth century onward, this physicalist vision amputated any concept of reality that could not be described in geometric terms.[4] Once "this physical space was itself extended to infinity, there was no 'room' left for any kind of spiritual space."[5] Victoria Nelson writes:

> This drastic reinterpretation of reality as a material world in which one's only transcendental link to God is internal [interiorized], I would argue, marks the real dividing line in Western culture. The turning point was not, as usually believed, that between the ancient world and the Christian culture that grew out of it, nor that between the Middle Ages and the Renaissance, or that between preindustrial and industrial society, but rather between Western culture up to the seventeenth century and what came after.[6]

Nelson also believes that the scientific developments and, somewhat later, the Industrial Revolution "conspired to produce a more profound shift in the Western worldview than anything in the preceding three millennia."[7]

Be that as it may, note Nelson's point about the development of a transcendental link to God that is only internal. In other words, the "same paradigm shift that banished both the underworld and the otherworld as 'real' locations" simultaneously opened up "possibilities of nontranscendental worlds other than our own."[8] One of these worlds is the introscopic realm or inner regions of the psyche, though many would believe that the psyche is itself transcendent, a sort of last frontier of reality. Thus for some the transcendent in Western societies became deeply identified with interiority and subjectivity.[9] After all, "once the physical world became infinite, where could any kind of spiritual realm possibly be?"[10] The answer, of course, is within the individual. Separating the individual from collectivities in fact encourages the interiorization of the individual since the transition to modernity was not so much a fragmenting of society into individual units, but rather a hollowing out of the individual, a mining of new psychogeographies. Campbell insightfully echoes this idea when he points out that Max Weber's disenchantment of the world required a parallel process: the "'enchantment' of the psychic inner world."[11]

In other words, the religious dualism of matter and spirit evolved into the Cartesian dualism of body and mind. The advent of the Enlightenment would result in the "annihilation of soul or spirit as categories of the real."[12] However, spiritual space did not only retreat into the mind; rather, through a strange metamorphosis, it greatly expanded the mental landscape of the individual. While telescopes, the calculus, and advances in cartography turned the macrocosm into macroscapes, the mapping of new geographies of consciousness turned its counterpart, the introcosm, into introscapes in which new territories of the mind were discovered. As the macrocosm of finite discontinuous spiritualized space was reduced to infinite continuous physical space, individuals lost their introcosmic souls but gained an introscopic inner world. Angels, demons, and gods migrated from the spiritual realm into the psychological realm, becoming ids, egos, superegos, executive egos, self, personality, cognitive processes, etc. The discipline that developed to account for this new mental realm, Psychology, would not emerge until the late nineteenth century.

I would like to conclude by taking a quick look at the future by suggesting that in the same way that consumerist desire, capitalism, and technology configure definitions of real space, these same forces are producing even newer imaginary geographies. Such spaces impact configurations of mind/body duality.

For example, in *The Pearly Gates of Cyberspace: A History of Space from Dante to the Internet* (1999), Margaret Wertheim retraces the development of spatial conceptions and argues that the newest type of space, cyberspace, represents a modern version of Medieval "soul space." Wertheim contends that the invention of Purgatory in 1274 by the Council of Lyons—a space that was neither Heaven nor Hell, but a special place for spiritual atonement—has "important parallels with the creation of cyberspace today."[13] After all, the principle of the correspondences between macrocosm and microcosm—"as above, so below"—has not changed over the centuries. Physical and metaphorical spaces are interwoven in complex and subtle ways; not only does the psyche use physical space to generate metaphoric space, but also metaphoric spatial conceptions can alter our understandings of physical space. The relation between physical and metaphorical space is thus a two-way street since they readily exchange meanings. What we have is a folding of "real" spaces into imagined places, and a projection of interior visions onto the exterior world. The interchangeability of different types of spaces, or its transposable nature, allows us to appreciate how physical sites and psychological states, driven by technology, shape each other. Such transference encourages the merging of boundaries between real space and not-so-real space. In technological terms, we can see how writing, printing, and computing have augmented the interior life of the individual, constructing increasingly meta-framed and super-reflexive narrative selves. The cosmological stuff of modernity is one in which different forms of abstraction—spatial, temporal, mathematical, scientific, commercial—confirmed and reinforced each other. The intensification of interiority has spun introscopic sceneries into existence in which a unique self dwells that expresses its own individuality by pursuing consumerist desires. This expanded sphere of personal subjectivity not only requires nuanced exploration of emotional subtleties but is also filled with newer and newer social needs.[14] Over history the individual introcosm has become an introscopic field of complex dynamics, demonstrating that spatiality is not merely a physical raw material; it is also a *psychological resource* that will develop in response to emerging techno-economics.

NOTES

1. Heer, 1963, p. 375, 374.
2. Wertheim, 1999, pp. 149, 152.
3. Mumford, 1934, p. 20.
4. Eventually, as we know, science led to relativistic conception of space and mind-boggling notions of hyperspace.
5. Wertheim, 1999, p. 151.
6. Nelson, 2001, p. 56.
7. Nelson, 2001, p. 56.
8. Nelson, 2001, p. 8.
9. Indeed, the emergence of Romanticism may be described as the linking of the cosmic transcendent with personal interiority.
10. Wertheim, 1999, p. 151.
11. Campbell, 1987, pp. 72–73. See Weber, 1978.
12. Wertheim, 1999, p. 37. Denying the "inner life" has had a much bigger influence on Psychology than most probably think. Though certain traditions of research Psychology kept alive the idea of an interior psychoscape, other influential traditions, particularly American behaviorism, made strenuous efforts to deny the existence of internal mental mechanisms, or what we ordinarily call the mind. However, some Psychologists, neuroscientists, and the lay person have had a hard time believing that conscious subjective thought is unreal. Moreover, the last several decades have witnessed a readmittance of the mind into the research agenda of Psychology due to the so-called "cognitive revolution." Nevertheless, studies of "what goes on inside the head" are still configured by a powerful dualism of mind versus matter.
13. Wertheim, 1999, p. 69.
14. As an example of what I mean, see McVeigh, 2013, "Mental Imagery and Hallucinations as Adaptive Behavior: Divine Voices and Visions as Neuropsychological Vestiges." I attempt to demonstrate how divine visions (hallucinations or "extraceptions") and mental imagery (introceptions) are both "superceptions." These experiences, though a historico-cultural development, have functioned as adaptations to changing historical circumstances.

Appendix A

How to Open the "Black Box": Cultural Psychology

For most the proper place for the study of mind is in Psychology. However, speaking of the related field of cognitive science (a disciplinary merger of Psychology, linguistics, computer science, and philosophy of mind), Gardner lists what it de-emphasizes: emotions, historical factors, and the sociocultural context in which actions or thoughts transpire.[1] Thus, "stimuli, contexts, resources, values, meanings, knowledge, religion, rituals, language, technologies, [and] institutions" are all regarded as external to mind.[2] But this approach is idealized, limiting, and artificially bounded. The issue is not merely academic because how we conceive the relation between the mind and its environment has practical implications for how we approach social problems (e.g., psychotherapy, treatment of mental health, development of psychotropic drugs). Following Shweder, I delineate the meaning of cultural Psychology by explaining *what it is not.*

IT IS NOT GENERAL PSYCHOLOGY

Arguably conventional Psychology's primary goal is to describe a presupposed "central processing mechanism" (CPM) that is abstract, hidden, fixed, and universal. In their search for the Holy Grail of the CPM, many Psychologists have little concern for the concrete, particular stuff, substance, or variable content that "is operated upon by the processor or may interfere with its operations." In short, mainstream Psychology searches for the CPM's "invariant laws of operation" that are supposedly "untainted by content and context." A society's cosmology, doctrines, ideas, and values constitute the "particular stuff, substance, or content" which constitute a given cultural Psychology. The assumption that beliefs (culture) are something layered over deeper, pre-cultural mental processes (psychological structures) is misleading. No fundamental division between the CPM and an individual's historical context, institutional setting, beliefs, values, and knowledge exists.[3]

IT IS NOT CROSS-CULTURAL PSYCHOLOGY

A subdiscipline of general Psychology, cross-cultural Psychology, also attempts to "itemize the features of the CPM." General research Psychology has more faith in psychic unity than cultural Psychology, which has a "fundamental skepticism concerning all those fateful and presupposed distinctions between" mind and environment, form and content, the deep and the superficial.[4] I am not interested in peeling back cultural accretions layered over a CPM. Such mechanisms, of course, certainly exist, but I highlight not psychic unity but psychic plasticity and diversity. Consider the practical implications of what we have discussed so far:

Psychic Plasticity during One's Life Course

The psyche changes over an individual's lifecycle, from the time an infant is born until a person dies. During this period, new experiences and information alter an individual's psyche, resulting in an endless series of socializations, desocializations, and resocializations.

Psychic Plasticity across Different Cultures

Due to socially pragmatic reasons, different communities possess dissimilar ethno-mentalities. These communities may be apparent at the microlevel: families, dyads, small groups, small-scale communities, occupational situations, or at the macrolevel: ethnic groups, religious communities, nations, states, etc.

Psychic Plasticity through History

Different historical periods are characterized by distinct ethno-mentalities. Human nature and our thinking apparatus are not an ahistorical constant: People are adapted to "change and constitutionally equipped with organs which enable them to learn constantly . . . and to change the pattern of their social life together."[5] To quote: Human nature is "fundamentally and radically malleable. It should not at all be understood as an all-inclusive immutable state which always remains the same. As such, there is no bench-mark of some universal human condition."[6]

IT IS NOT PSYCHOLOGICAL ANTHROPOLOGY

Some researchers, such as Psychological anthropologists, have attempted to utilize the "stuff of culture to characterize or discover" a CPM. But Psychological anthropology suffers from reductionism because it assumes that society is configured by the projection of dark psychic forces, "operated upon by, or expressive of, deep and invariant psychological laws or processes or motivation, affect and intellect." So Psychological anthropology, like general Psychology, does not concern itself with how the psyche is "fundamentally altered by the content, stuff, material, or sociocultural environment on which it operates."[7]

IT IS NOT ETHNOPSYCHOLOGY

Cultural Psychology is related to a subdiscipline of ethnosemantics (or ethnoscience) called ethnopsychology. However, the latter investigates mind, self, and emotions in the same way it approaches folk beliefs about botany or kinship—i.e., it is not person-centered enough. Cultural Psychology gives more attention to the experiences of actual persons.[8] Note how Catherine Lutz, in her investigations of emotions, contends that they "can more profitably be viewed as serving complex communicative, moral, and cultural purposes rather than as labels for internal states whose nature or essence is presumed to be universal."[9] Though she is specifically discussing emotions, her comments also apply to other psychological processes. In this book I address spatiality, the passing of time, conscious interiority as experienced by individuals. These are not readily catalogued epiphenomena à la ethnopsychology. For cultural Psychology, then, the psyche is "content-driven, domain-specific, and constructively stimulus bound; and it cannot be extricated from the historically variable and cross-culturally diverse intentional worlds in which it plays a coconstituting part."[10]

The aim, then, when exploring the historical trajectory of mind, is to "deconstruct an overly naturalized and rigidly bounded concept" and to "treat [it] as an ideological practice rather than as a thing to be discovered or an essence to be distilled."[11] Constructed in particular places and periods, the psyche cannot be reduced to computational processes immune from politics, economics, and other influences. The disciplinary divisions between formal individual psychological processes and sociocultural content reflect the demarcations between Psychology, sociology, anthropology, etc., not natural cleavages between the person and society.

The point of this aforementioned discussion was to provide some perspective for the following claim that informs my contentions: Cultural changes of the psyche are much more profound and influential than we have assumed. Indeed, throughout the centuries changes in the shape of the putty (rather than the form of the vase) constitute major adaptive leaps. In other words, the functional impact of neurological evolution and cultural psychological modifications are arguably not that different.

NOTES

1. Gardner, 1985, p. 6.
2. Shweder, 1990, p. 5.
3. Shweder, 1990, pp. 4–7.
4. Shweder, 1990, pp. 9, 13.
5. Elias, 1978, p. 115.
6. Tester, 1992, p. 86.
7. Shweder, 1990, pp. 14, 15–16.
8. Shweder, 1990, p. 16.
9. Lutz, 1988, p. 5.
10. Shweder, 1990, p. 13.
11. Lutz, 1988, p. 4.

Appendix B

What Conscious Interiority Is Not

So much fuzzy thinking and confusion surrounds the nature of conscious interiority that it is necessary to first clarify what interiority *is not.* This is the purpose of this appendix.

INTERIORITY NOT A COPY OF EXPERIENCE

The most common and misleading understanding of what constitutes interiorized experience is that it is perception, or that introspection is some type of mental mirror that "copies" what we perceive. The perception theory of conscious interiority is the most obstinate, commonly-held, and entrenched views in Psychology. This theory is not sort of wrong; rather, it is wrong in the most fundamental sense. The belief that interiority is perceptual experience is very commonsensical, but the confusing of introspectable consciousness with what we sense has puzzled thinkers for centuries. It is, however, very easy to demonstrate that the two are different, as the following sections argue. That interiority does not reflect the outside world is clear: what we can recall is a mere "thimbleful" in the huge oceans of our actual knowledge.[1] Our introspection only reflects a very limited portion of our perceptual, bodily experiences. Moreover, people are "not conscious of very much of what they sense; people are not conscious of very much of what they think; people are not conscious of very much of what they do." Conscious interiority, then, "is a closed, partly mendacious system," which fools into us believing that what we experience is what we perceive.[2]

The very language we use to describe psycho-physiological processes requires an overhaul. Let us avoid confusion by labeling nonconscious cognition "reactivity" (rather than the confusing "perception"). Reactivity includes "all stimuli my behavior takes into account of in any way, while consciousness is something quite distinct and a far less ubiquitous phenomenon. We are conscious of what we are reacting to only from time to time."[3] Remarkably, an "astonishing number" of physiology and neuropsychology textbooks fail to mention this.[4] However, "much of everyday life—thinking, feeling, and doing—is automatic in that it is driven by current features of the environment (i.e., people, objects, behaviors of others, settings, roles, norms, etc.) as mediated by automatic cognitive processing of those features, without any mediation by conscious choice or reflection."[5] Note that "over eleven million bits a second from the world" moves through our sensory mechanisms. However, the physiologist Dietrich Trickner reports that "only *one millionth* of what our eyes see, our ears hear, and our other senses inform us about *appears in our consciousness.*"[6] The task of psyche is to "reduce eleven million bits to sixteen bits so that the sixteen bits can be used as a map of the eleven million."[7] According to Professor Manfred Zimmerman of Heidelberg University, what we perceive at any moment is "limited to an extremely small compartment in the stream of information about our surroundings floating inform the sense organs." Research has demonstrated that the "maximal information flow of the process of conscious sensory perception is about 40 bits . . . many orders of magnitude below that taken in by receptors." Our perception, then, would appear to be limited to a minute part of the abundance of information flowing in from the senses.[8] Remarkably, though Professor Zimmerman's conclusion has been repeated since the late 1950s in the literature on medicine, Psychology, and information theory, it has not made "a mark either on physiology and neuropsychology or on our culture as whole."[9]

An example of reactivity is proprioception, or the largely nonconscious sense of where one's body parts are in space (its impairment is related to the loss of "body-sense" and "body-ego"). An understanding of reactivity allows us to see through the illusions of conscious interiority. "We experience sensation but do not experience that this sensation has been interpreted and processed." The processing is the task of the nonconsciousness. People experience far more than what they consciously register; they "interact far more with the world and with each other than their consciousness thinks they do," and the "control of actions that consciousness feels it exercises is an illusion."[10] "Few hypotheses in the behavioral sciences have occasioned so much controversy as the suggestion that people may be affected by external stimuli of which they remain wholly unaware."[11] Why? Because we are socialized to believe that we are autonomous, independent, and self-directed; the self is sovereign.

THE ILLUSION OF ALWAYS BEING CONSCIOUS

Much of the day we are not conscious, and most of our actions are habitual and scripted. However, the notion of habit immediately conjures up behavior that "consists in a fixed, mechanical reaction to particular stimuli and is, as such, devoid of meaning from the actor's point of view." Indeed, "contemporary sociology has virtually dispensed with the concept." Pointing to the work of the "father of sociology," Émile Durkheim, Camic writes that there "was the idea that, by its very nature, human action, whether individual or collective, oscillates between two poles, that of consciousness or reflection on the one side, and that of habit on the other side, with the latter pole being the stronger." We like to believe that human action is "purposive, rational, voluntaristic, or decisional."[12]

The Overlooked Importance of Habitual Behavior

Our belief that our "unthinking" behavior is mere reaction or uncomplicated is a grave misperception because habit possesses intentionality and has behind it the entire mental machinery of the individual. Information of the concrete and specific details needed to carry out an act sinks into the nonconscious as one becomes more familiar with a skill (e.g., learning to type and driving a car are oft-cited examples). "Preconscious automaticity models the regularities in one's reaction to an event, and eventually subsumes them so that the conscious mind no longer has to make decisions and understandings it always makes the same way anyway."[13] Interiority arises as a "state of relative indeterminacy" and then breaks the link in the causal chain that usually leads straight from motivations and triggering cues to behavior.[14]

As a cognitive process, then, conscious interiority usually kicks in when our environment confronts us with novel problems that our preprogrammed and scripted behavior cannot adequately handle. We live in "immediate acts of experience"; it is only when our actions are in some way frustrated that awareness of self and our motives become apparent.[15] The hindering of unmindful and routinized activity (reactivity) often causes a rupture between spontaneous and more directed forms of self-presentation. This is why "under some circumstances in everyday life the actor becomes, is, or is made *aware* of an actual or potential discrepancy between his 'real' and his 'projected' selves, between his 'self' and his 'character.'"[16]

Scripted behavior gives us the illusion that we must be interiorizing most of the time. In fact, interiority is a much smaller part of our mental life than we are aware of,

> because we cannot be conscious of what we are not conscious of. How simple that is to say; how difficult to appreciate! It is like asking a flashlight in a dark

> room to search around for something that does not have any light shining upon it. The flashlight, since there is light in whatever direction it turns, would have to conclude that there is light everywhere. And so consciousness can seem to pervade all mentality when actually it does not.[17]

In other words, the "criterion for consciousness is quite simply consciousness. If you are not conscious of something, you are not conscious of it."[18]

THE ILLUSION THAT INTERIORITY REFLECTS COGNITION

Consider the koanlike statement that being aware of an experience means that it has passed (or that conscious interiority operates far slower than nonconscious cognition). Or that what we experience "is a lie, for we experience it as if we experienced it before we experienced it." However, "there is a good point in this fraud, because what we need to know is when our skin was pricked, not when we became conscious of it." The point here is that nonconscious cognition processes occur rapidly and in parallel, while the conscious processes of interiority are "slower and serial—they take one thing at a time, like an old-fashioned computer." Interiority lags behind our reactions to the environment. It takes half a second to become conscious of something; however, that it is not how we experience it. In other words a "subjective relocation in time" occurs; interiority "performs a temporal readjustment backward."[19]

The Speed of Thought

To understand how interiority performs a temporal readjustment backward, the actual speed of nonconscious and conscious cognition needs to be appreciated. Research conducted by H. Kornhuber et al. in 1976 demonstrated that the "readiness potential" (*Bereitschaftspoential*) starts "0.55 second before the act, while consciousness [interiority] starts 0.20 second before the act. The conscious decision thus takes place 0.35 second *after* the readiness potential commences."[20] Karl Küpfmüller argued that the capacity or "bandwidth" of conscious interiority is fifty bits a second of information.[21] He estimated that for the activity of silent reading, conscious interiority processes information at the rate of forty-five bits per second; for counting objects, three bits per second. Helmuth Frank estimated that it was sixteen bits a second, and came up with a "subjective time quantum (*subjektives Zeitquant*), which denotes a "psychological moment."[22] This is defined as the "smallest space of time we can experience the temporal resolution ability in human perception."[23]

Nørretranders points out that research into the capacity of consciousness ran out of steam in the 1960s; "it is a mystery why the field has been allowed to die out." He asks, "Why was the revolutionary view of man invited by these results so completely ignored?" He answers that the idea that we can be influenced by perceptions that do not result in conscious interiority but are still somehow lodged within the individual has always been associated with "considerable fear." Perhaps such research challenged the view that the self is autonomous and sovereign. Nørretranders notes that interest in subliminal perception rose in the late 1800s, reached a zenith around 1960, abruptly declined, and since then has been gradually increasing.[24] If anything, all this clearly demonstrates how research agendas are shaped by what we *prefer* rather than what we *should* know.

INTERIORITY NOT NECESSARY FOR THINKING OR REASONING

Thinking is "an automatic process following a struction [from material *constructions* for and *instructions* to one's mind] and the "materials on which the struction is to operate."[25] Indeed, interiority is not only unnecessary for many psychological processes; it is often counterproductive.[26] This is why many skills are best carried out when we do not think about what we are doing.[27] Certainly the reader has had the experience of almost falling down a staircase when trying to be conscious of each step, or mangling a sentence on a computer screen when rapidly typing but selectively thinking of each key to be hit. We "may be aware of the goals of and the conditions or procedures and the products of their executions"; but we are unaware of the operations themselves.[28] There is "insufficient evidence to support the position that conscious mediation of situational effects is the rule rather than the exception." Therefore, the "assumption of conscious mediation" deserves the "same scientific scrutiny as the assumption of automaticity."[29] Also, interiority is not necessary for concepts, which merely categorizes behaviorally equivalent things. Interiority is "not only not the repository of concepts; it does not usually work with them at all!"[30] For many of us this is a most counterintuitive assertion.

However, our "minds work much faster than consciousness can keep up with. We commonly make general assertions based on our past experiences in an automatic way, and only as an afterthought are we sometimes able to retrieve any of the past experiences on which an assertion is based." As for creativity, it requires several stages: preparation, incubation, and illumination (the latter is justified by logic after the fact). The "period of preparation is essentially the setting up of a complex struction together with conscious attention to the materials on which the struction is to work. But then the actual process of reasoning, the dark leap into huge discovery . . . has no representation in consciousness. Indeed, it is sometimes almost as if the problem had to be forgotten to be solved." And consider why we need logic: "Reasoning and logic are to each other as health is to medicine. . . . Reasoning refers to a gamut of natural thought processes in the everyday world. Logic is how we ought to think." Hence we need logic, a type of meta-framing, because most reasoning is not conscious.[31]

Related to the myth that "thinking is conscious" is another mistaken view inspired by an Enlightenment bias: the myth of "rationality," or that the mind is accessible and displays its contents for a transcendental "I" (of a Cartesian–Kantian lineage) to view. Passions, strong sentiments, and intruding thoughts, often associated with nonconscious impulses, merely cloud the mirror of reason. Conscious thought is rational, nonemotional, and logical, while nonconsciousness is irrational, emotional, and illogical (note the influence of Freud here). Most mainstream researchers used to regard the nonconscious as secondary, subsidiary, and subordinate to the more primary, dominant, and controlling conscious mind. However, much recent research has conclusively demonstrated that nonconscious thought patterns perform a primary role in our social life. One of the essential functions of psyche is to trick us, to mask its operations. We do not necessarily know why we do what we do, as recent work on nonconscious cognition has demonstrated. The purpose of consciousness is *not* to reveal the inner workings of the mind; rather, it is to hide cognitive functioning, to fool us that inner workings of the mind do not exist. The real purpose of conscious interiorization is to assist us in taking shortcuts, to navigate through the world by offering us a menu of options that we can visualize "in" our psychoscapes.

INTERIORITY IS NOT NECESSARY FOR LEARNING

Our common folk Psychology, without careful consideration, has inherited the idea that consciousness is necessary for learning. Interiority brings us the task, providing us with the goal to be attained. From then on, however, "apart perhaps from fleeting neurotic concerns about your abilities at such tasks," the learning happens for us.[32] Only the preparation, materials, and end result of learning are consciously interiorized. Interiority, then, is not necessary for learning, and in fact, often gets in the way of learning acquiring new skills.

DECISIONS ARE NOT INITIATED BY A CONSCIOUS SELF

The self is an object of the conscious introspect-able mind, metaphorically "seen" by the "mind's eye." As such, it is a product of complex nonconscious cognitive processes, and though commonsensical Psychology imagines the self as an executive ego, director, central processing mechanism, or homunculus that has decision-making powers over the body—like a puppeteer pulling the strings—such a view is an ideological construct. The self is not a decision-making agent; rather, it is the product of many decision-making processes that occur unawares. This is not to deny agency to individuals, but to argue that human decision-making is intricate, compounded, multiple, and for the most part, invisible. In other words, the "subjective unity" of self, thought, and of personal experience is illusory.[33] Conscious interiority

> tells us that we can decide on what we do. Yet it is apparently a mere ripple on the surface, a little tin god pretending to be in charge of things beyond its control. Our consciousness claims that it makes the decisions, that it is the cause of what we do. But our consciousness is not even there when the decision is made. It lags behind, but it does not tell us that. It dupes itself—but how can my consciousness dupe itself without duping me? Is the self-duplicity of consciousness not my own self-duplicity?[34]

Consciousness does not initiate actions, but it does decide which actions should not be carried out. In other words, the purpose of conscious interiority is to provide "the veto"—i.e., one can stop an impulse: vetoes appear frequently only in circumstances that are themselves infrequent.[35] Baumeister and Sommer propose that the role of consciousness is to "override automatic, habitual, or standard responses on the infrequent occasions when such intervention is needed. Consciousness thus undermines the lawful, predictable nature of human behavior and produces a situation of relative indeterminacy." Thus, conscious interiority "creates a hole or gap in the deterministic web of causal relationships that shapes human behavior, so to speak. Under familiar, comfortable circumstances, certain causes lead smoothly to certain behavioral responses. Consciousness can disrupt and alter these connections, thereby disengaging behavior from its usual causes."[36] Or we might say that processes implicated in individual responsibility and free will function not to initiate voluntary acts but to "select and control volitional outcomes."[37] After all, we

> feel most content when our consciousness does not exercise free will. . . . People are happiest when it is not their consciousness that selects the nonconscious urges to act. People feel most content when they just act. . . . But the consequence of this observation is that we must face the fact that it is not our consciousness that is in charge when we're feeling good. . . . So we must ask: Do we possess free will only when we are feeling bad? Or do we also possess free will when we feel good? And if so, who possesses it?[38]

NOTES

1. Jaynes, 1976, p. 28.
2. Nørretranders, 1998, pp. 259, 209.

3. Jaynes, 1976, p. 22. "And whereas reactivity can be defined behaviorally and neurologically, consciousness at the present state of knowledge cannot" (Jaynes, 1976, p. 22).
4. Nørretranders, 1998, p. 124.
5. Bargh, 1997, p. 2.
6. Trickner, 1966, p. 6, emphasis in original; cited in Nørretranders, 1998, p. 126.
7. Nørretranders, 1998, p. 242.
8. Zimmerman, 1989, pp. 172. See also Zimmerman, 1986, p. 116.
9. Nørretranders, 1998, p. 125.
10. Nørretranders, 1998, pp. 288, iv.
11. Dixon, 1971, p. 5.
12. Camic, 1986, pp. 1046, 1052, 1040.
13. Bargh, 1997, p. 10. Bargh distinguishes between preconscious and postconscious forms of automaticity (1997, p. 3); "Preconscious processing is that initial stage of cognition in which the world makes contact with our mind" (Bargh, 1997, p. 8). Also, for a review of the automaticity literature, see Bargh, 1994. See also Bryan and Harter, 1897, 1899; James, 1890; LeBarge and Samuels, 1974; Schneider and Shiffrin, 1977; Solomons and Stein, 1886; Posner and Snyder, 1975; Vera and Simon, 1993; and Wyer, 1997.
14. Baumeister and Sommer, 1997, pp. 79, 77.
15. Mill, 1984, p. 208.
16. Messinger, Sampson, and Towne, 1990, p. 74, emphasis in original.
17. Jaynes, 1976, p. 23.
18. Nørretranders, 1998, p. 220.
19. Nørretranders, 1998, pp. 128, 165, 129, xi, 235. Interestingly, conscious interior experience *"is projected back in time in exactly the same way as a stimulation of the sensory cortex is projected out onto the body"* (Nørretranders, 1998, p. 235, emphasis in original).
20. Nørretranders, 1998, pp. 218–19, emphasis in original. Apparently, stimuli shorter than half a second are not experienced at all (Nørretranders, 1998, p. 227).
21. Küpfmüller, 1962, p. 1500; see also Küpfmüller, 1971.
22. Frank, 1962, p. 69.
23. Nørretranders, 1998, p. 139. See also the research by S. Lehrl and B. Fischer, 1985, and Richard Jung, "Perception, Consciousness and Visual Attention," 1978.
24. Nørretranders, 1998, pp. 141, 157, 158.
25. Jaynes, 1976, p. 39.
26. Baumeister and Sommer, 1997, p. 76.
27. Nørretranders, 1998, p. 141.
28. Khilstrom, 1987, p. 1446.
29. Bargh, 1997, p. 5.
30. Jaynes, 1976, p. 31.
31. Jaynes, 1976, pp. 42, 44, 41.
32. Jaynes, 1976, pp. 31, 33.
33. Oakley and Eames, 1985, p. 247.
34. Nørretranders, 1998, p. 220.
35. Nørretranders, 1998, pp. 243, 248.
36. Baumeister and Sommer, 1997, pp. 75, 79.
37. Libet, 1985a, p. 538. See also Libet 1965, 1981, 1989, 1990, 1991, 1992; Libet et al. 1964, 1967, 1972, 1979, 1983, and 1991.
38. Nørretranders, 1998, p. 250.

Appendix C

Spaces: Real and Imaginary

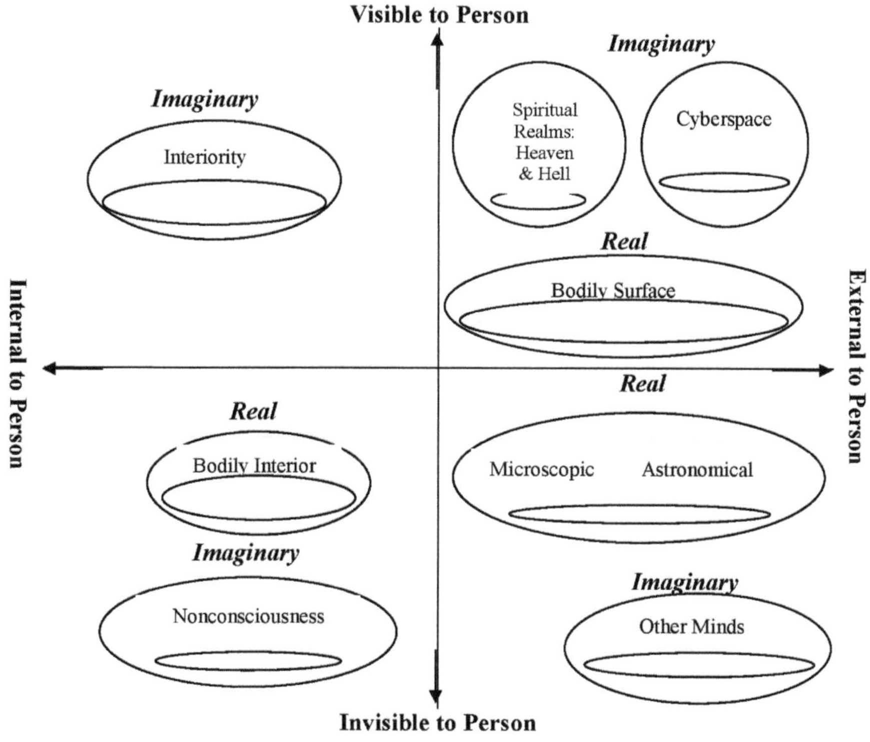

Appendix D

The Visible, Invisible, and Introspectable

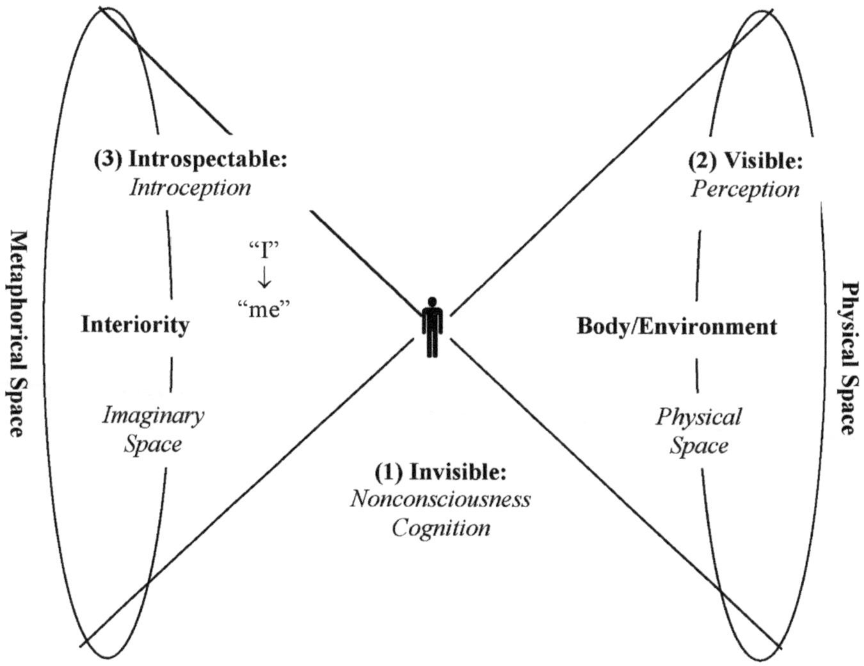

References

Adorno, T. W., Else Frenkel-Brunswik, Daniel J. Levinson, and R. Nevitt Sanford. 1950. *The Authoritarian Personality*. New York, NY: W. W. Norton.

Aquinas, Thomas. 1955–57. *Summa contra Gentiles*. Translated by Anton C. Pegis et al. New York: Doubleday.

———. 1963–75. *Summa Theologiae* (60 vols.). Ed. Thomas Gilby et al. New York: McGraw.

Augustine of Hippo. 1984. *The City of God*. Ed. David Knowles. London: Penguin.

Azouvi, François. 2000. "*Physique* and *Moral*." In *Psyche and Soma*, ed. John P. Wright and Paul Potter, 267–80. Oxford: Clarendon Press.

Bain, Alexander. 2000 (1855). *The Senses and the Intellect*. Whitefish, MT: Kessinger Publishing.

———. 2004 (1872). *Mind and Body: The Theories of Their Relation*. Whitefish, MT: Kessinger Publishing.

———. 2007 (1859). *The Emotions and the Will*. University of Michigan Library: Scholarly Publishing Office.

Barclay, Michael W. 1997. "The Metaphoric Foundation of Literal Language: Towards a Theory of the Reification and Meaning of Psychological Constructs." *Theory & Psychology*, 7(3):355–72.

Barfield, Owen. 1967. *History in English Words*. Grand Rapids, MI: William B. EerdmansPublishing Co.

Bargh, J. A. 1982. "Attention and Automaticity in the Processing of Self-relevant Information." *Journal of Personality and Social Psychology*, 43:425–36.

———. 1990. "Auto-motives: Preconscious Determinants of Thought and Behavior." In *Handbook of Motivation and Cognition* (2 vols.), ed. E. T. Higgins and R. M. Sorrentino, 93–130. New York: Guilford.

———. 1992. "Being Unaware of the Stimulus versus Unaware of Its Interpretation: Why Subliminity per se Does Not Matter to Social Psychology." In *Perception without Awareness*, ed. R. Bornstein and T. Pittman, 236–55. New York: Guilford.

———. 1994. "The Four Horsemen of Automaticity: Awareness, Intention, Efficiency, Control in Social Cognition." In *Handbook for Social Cognition* (2nd ed.), ed. R. S. Wyer and T. K. Krull, 1–44. Hillsdale, NJ: Lawrence Erlbaum Associates.

———. 1996. "Principles of Automaticity." In *Social Psychology: Handbook of Basic Principles*, ed. E. T. Higgins and A. Kruglanski, 169–83. New York: Guilford.

———. 1997. "The Automaticity of Everyday Life." In *The Automaticity of Everyday Life*, ed. Robert S. Wyer, Jr., 1–62. Mahwah, NJ: Lawrence Erlbaum Associates.

———. 1989. "Conditional Automaticity: Varieties of Automatic Influence in Social Perception and Cognition." In *Unintended Thought*, J. S. Uleman and J. A. Bargh, 3–51. New York: Guilford.
Barnes, Jonathan, ed. 1984. *The Complete Works of Aristotle* (2 vols.). Princeton: Princeton University Press.
Baron-Cohen, Simon. 1995. *Mindblindness: An Essay of Autism and Theory of Mind*. Cambridge, MA: MIT Press.
Baumeister, Roy F., and Kristin L. Sommer. 1997. "Consciousness, Free Choice, and Automaticity." In *The Automaticity of Everyday Life*, ed. Robert S. Wyer, Jr., 75–81. Mahwah, NJ: Lawrence Erlbaum Associates.
Bayly, Christopher. 2004. *The Birth of the Modern World 1780–1914: Global Connections and Comparisons*. Oxford, UK: Blackwell.
Bell, Charles. 2004 (1837). *The Hand, Its Mechanism and Vital Endowments as Evincing Design*. Boston, MA: Adamant Media Corporation.
———. 2005 (1830). *The Nervous System of the Human Body*. Delran, NJ: Gryphon Editions.
———. 2006 (1872). *The Anatomy and Philosophy of Expression as Connected with the Fine Arts*. Whitefish, MT: Kessinger Publishers.
Benton, John F. 1982. "Consciousness of Self and Perceptions of Individuality." In *Renaissance and Renewal in the Twelfth Century*, ed. Robert L. Benson and Giles Constable, 263–95. Cambridge, MA: Harvard University Press.
Billow, R. M. 1977. "Metaphor: A Review of the Psychological Literature." *Psychological Bulletin*, 84:81–94.
Boorstin, Daniel J. 1985. *The Discoverers*. New York: Vintage Books.
Boring, E. G. 1929. *A History of Experimental Psychology*. New York: Appleton-Century-Crofts.
Brown. H. I. 1977. *Perception, Theory and Commitment: The New Philosophy of Science*. Chicago: University of Chicago Press.
Brožek, Josef M. 1999. "From 'Psichiologia' to 'Psychologia': A Graphically Documented Archival Study across Three Centuries." *Journal of the History of the Behavioral Sciences*, 35(2): 177–80.
Bryan, W. L., and N. Harter. 1897. "Studies in the Physiology and Psychology of Telegraphic Language." *Psychological Review*, 4:27–53.
———. 1899. "Studies of the Telegraphic Language. The Acquisition of a Hierarchy of Habits." *Psychological Review*, 6:345–75.
Bury, J. B. 1932. *The Idea of Progress*. New York: Macmillan.
Camic, Charles. 1986. "The Matter of Habit." *American Journal of Sociology*, 91(5):1039–87.
Campbell, Colin. 1987. *The Romantic Ethic and the Spirit of Modern Consumerism*. Oxford: Basil Blackwell.
Childe, V. Gordon. 1951. *Man Makes Himself*. New York: New American Library.
Comte, Auguste. 1835–52. *Cours de philosophie positive* (6 vols.). Paris: Bachelier.
Condorcet, Marie Jean Antoine Nicolas Caritat. 1955 (1795). *Sketch for a Historical Picture of the Progress of the Human Mind*. Translated by June Barraclough. London: Weidenfeld and Nicolson.
Cornford, Francis MacDonald, trans. 1968. *The Republic of Plato*. London: Oxford University Press.
Cornish, Edward. 1977. *The Study of the Future*. Washington, DC: World Future Society.
Cox, Harvey. 1969. "Christianity's Conflicting Views of the Future." *The Futurist*, 3(4) (August): 95–96.
Crary, Jonathan. 1994. *Techniques of the Observer: On Vision and Modernity in the Nineteenth Century*. Cambridge, MA: MIT Press.
Crawley, A. E. 1909. *The Idea of the Soul*. London: Adam and Charles Black.
Crosby, Alfred W. 1997. *The Measure of Reality: Quantification and Western Society, 1250–1600*. Cambridge University Press.
Danziger, Kurt. 1994. *Constructing the Subject: Historical Origins of Psychological Research*. New York: Cambridge University Press.

Darwin, Charles. 1859. *The Origin of Species by Means of Natural Selection, or the Preservation of Favored Races in the Struggle for Life.* London: John Murray.
Datson, Lorraine, and Peter Galison. 2007. *Objectivity.* New York: Zone Books.
Descartes, Rene. 1989 (1649). *The Passions of the Soul.* Translated by Stephen H. Voss. Indianapolis, IN: Hackett Publishing.
———. 1996 (1642). *Meditations on First Philosophy.* Translated by Joseph Cottingham. Cambridge: Cambridge University Press.
———. 2006 (1633). *The World and Other Writings.* Ed. Stephen Gaukroger. Cambridge: Cambridge University Press.
———. 2006 (1637). *Discourse on Method.* London: Pomona Press.
Dixon, Norman. 1971. *Subliminal Perception: The Nature of a Controversy.* London: McGraw Hill.
Duchesneau, François. 2000. "Stahl, Leibniz, and the Territories of Soul and Body." In *Psyche and Soma*, ed. John P. Wright and Paul Potter, 217–36. Oxford: Clarendon Press.
Durkheim, Emile. 1964 (1893). *The Division of Labor in Society.* Translated by George Simpson. New York: Collier Books.
Edgerton, 2009. *The Mirror, the Window, and the Telescope: How Renaissance Linear Perspective Changed Our Vision of the Universe.* Ithaca, NY: Cornell University Press.
Elias, Norbert. 1969 (1939). *The Civilizing Process, Vol. I. The History of Manners.* Oxford: Blackwell.
———. 1978. *What Is Sociology* (trans. Stephen Mennell and Grace Morrissey). New York: Columbia University Press.
———. 1982 (1939). *The Civilizing Process, Vol. II. State Formation and Civilization.* Oxford: Blackwell.
Ellenberger, H. F. 1970. *The Discovery of the Unconscious.* New York: Basic Books.
Engels, Friedrich. 1942 (1884). *The Origin of the Family, Private Property and the State: In the Light of the Researches of Lewis H. Morgan.* New York: International Publishers.
Evreinoff, Nicholas. 1990. "Appendix II: The Never Ending Show." In *Life as Theater: A Dramaturgical Source Book* (2nd edition), ed. Dennis Brissett and Charles Edgley, 419–25. New York: Aldine de Gruyter.
Fechner, Gustav Theodor. 1851. *Zend-Avesta, oder über die Dinge des Himmels und des Jensets.* Leipzig: Voß.
———. 1860. *Elemente der Psychophysik.* Leipzig: Breitkopf & Härtel.
———. 1876. *Vorschule der Aesthetik.* Leipzig: Breitkopf & Härtel.
Ferreiro, E. 1985. "Literacy Development: A Psychogenetic Perspective." In *Literacy, Language, and Learning: The Nature and Consequences of Reading and Writing*, ed. D. R. Olson, N. Torrance, and A. Hildyard, 217–28. New York: Cambridge University Press.
———. 1991. "Psychological and Epistemological Problems on Written Representation of Language." In *Learning and Instruction: European Research in an International Context* (vol. 3), ed. M. Carretero, M. Pope, R. J. Simons, and J. Pozo, 157–73. Oxford: Pergamon Press.
Foucault, M. 1979. *Discipline and Punish: The Birth of the Prison.* New York: Vintage Books.
Frank, David John, John W. Meyer, and David Miyahara. 1995. "The Individualist Polity and the Prevalence of Professionalized Psychology: A Cross-National Study." *American Sociological Review*, 60(3):360–77.
Frank, Helmuth G. 1962. *Kybernetischen Grundlagen der Pädagoik* (2 vols.). Baden-Baden: Agis-Agis Verlag.
Frazer, James. 1890. *The Golden Bough: A Study in Magic and Religion* (2 vols.). London: Macmillan Publishing Company.
Freud, Sigmund. 1961 (1930). *Civilization and Its Discontents.* Edited and translated by James Strachey. New York: W. W. Norton and Company.
Fuchs, Alfred H. 2000. Contributions of American Mental Philosophers to Psychology in the United States. *History of Psychology* 3(1), 3–19.
Furedi, Frank. 2004. *Therapy Culture: Cultivating Vulnerability in an Uncertain Age.* London: Routledge.

Gardner, Howard. 1985. *The Mind's New Science: A History of Cognitive Revolution*. New York, NY: Basic Books.

Gellner, Ernest. 1983. *Nations and Nationalism*. Ithaca, NY: Cornell University Press.

Gendlin, Eugene. 1997. *Experiencing and the Creation of Meaning: A Philosophical and Psychological Approach to the Subjective*. Evanston, IL: Northwestern University Press.

Gentner, Dedre, and Jonathan Grudin. 1985. "The Evolution of Mental Metaphors in Psychology." *American Psychologist*, 20(4): 181–92.

Gibbs, R. 1994. *The Poetics of Mind: Figurative Thought, Language, and Understanding*. New York and Cambridge: Cambridge University Press.

———. 1999. "Taking Metaphor out of Our Heads and Putting It into the Cultural World." In *Metaphor in Cognitive Linguistics*, ed. R. Gibbs & G. Steen, 145–66. Amsterdam: John Benjamins.

———. 2003. "Embodied Experience and Linguistic Meaning." *Brain and Language*, 84:1–15.

Gibbs, R., and G. Steen, eds. 1999. *Metaphor in Cognitive Linguistics*. Amsterdam: John Benjamins.

Goffman, Erving. 1959. *The Presentation of Self in Everyday Life*. New York: Anchor Books.

Gross, Martin L. 1978. *The Psychological Society: A Critical Analysis of Psychiatry, Psychotherapy, Psychoanalysis and the Psychological Revolution. New York: Simon & Schuster*

Groves, Ernest R. 1933. *Marriage*. New York, NY: Henry Holt and Company.

Groves, Ernest R., and Phyllis Blanchard. 1930. *Introduction to Mental Hygiene*. New York, NY: Henry Holt & Company.

Grube, G. M. A., trans. 1992. *Plato: Republic*. Revised by C. D. C. Reeve. Cambridge: Hackett Publishing Co.

Gundert, Beate. 2000. "Soma and Psyche in Hippocratic Medicine." In *Psyche and Soma*, ed. John P. Wright and Paul Potter, 13–35. Oxford: Clarendon Press.

Harris, Marvin. 1968. *The Rise of Anthropological Theory*. New York: Crowell.

———. 1977. *Cannibals and Kings: The Origins of Cultures*. New York: Random House.

———. 1979. *Cultural Materialism: The Struggle for a Science of Culture*. New York: Vintage Books.

Hartley, David. 1966 (1749). *Observations on Man: His Frame, His Duty, and His Expectations*. New York: Scholars' Facsimiles & Reprints.

Haskell, Robert E., ed. 1987. *Cognition and Symbolic Structure: The Psychology of Metaphoric Transformation*. Norwood, NJ: Ablex Publishing.

Heckel, Theo K. 2000. "Body and Soul in Saint Paul." In *Psyche and Soma*, ed. John P. Wright and Paul Potter, 117–32. Oxford: Clarendon Press.

Heer, Friedrich. 1963. *The Medieval World*. New York: Mentor books.

Hegel, Georg Wilhelm Friedrich. 1956 (1837). *Lectures on the Philosophy of History*. Translated by J. Sibree. New York: Dover Publications.

———. 1967 (1803). *The Phenomenology of Spirit*. Translated by J. B. Baille. New York: Harper and Row.

Helmholtz, Hermann Ludwig Ferdinand von. 1881. *Popular Lectures on Scientific Subjects*. New York: Longmans, Green and Co.

———. 1962 (1856–67). *Treatise on Physiological Optics*, 3 vols. Ed. James P. C. Southall. Mineola, NY: Dover.

———. 1863 (1954). *On the Sensations of Tone*. Mineola, NY: Dover Publications.

Herbart, Johann Friedrich. 1891 (1816). *Textbook in Psychology: An Attempt to Found the Science of Psychology on Experience, Metaphysics, and Mathematics*. Translated by Margaret K. Smith. New York: Appleton.

Hobbes, Thomas. 1972 (1658). *De homine (On the Nature of Man)*. Garden City, NY: Anchor Books.

———. 1994 (1651). *Leviathan*. Ed. R. Tuck. Cambridge: Cambridge University Press.

Hume, David. 1955 (1748). *An Inquiry Concerning Human Understanding*. Ed. Charles W. Hendell. New York: Bobbs-Merrill.

Illouz, Eva. 2008. *Saving the Modern Soul: Therapy, Emotions, and the Culture of Self-Help*. Berkeley: University of California Press.

James, William. 1890. *Principles of Psychology* (2 vols.). New York: Holt.

Jaynes, Julian. 1976. *The Origin of Consciousness in the Breakdown of the Bicameral Mind.* Boston, MA: Houghton Mifflin Co.

———. 1990. "Afterword." In *The Origin of Consciousness in the Breakdown of the Bicameral Mind,* 447–69. Boston: Houghton Mifflin Co.

———. 1991. "Afterword." *The Origin of Consciousness in the Breakdown of the Bicameral Mind* (2nd edition). Boston, MA: Houghton Mifflin Co.

———. 2012. "A Two-Tiered Theory of Emotions." In *The Julian Jaynes Collection: Biography, Articles, Lectures, Interviews, Discussion,* ed. Marcel Kuijsten, 147–51. Henderson, NV: Julian Jaynes Society.

Johnson, M. 1987. *The Body in the Mind: The Bodily Basis of Meaning, Imagination, and Reason.* Minneapolis: University of Minnesota Press.

Johnson, M., ed. 1981. *Philosophical Perspectives on Metaphor.* Minneapolis: University of Minnesota Press.

Jung, Richard. 1978. "Perception, Consciousness and Visual Attention." In *Cerebral Correlates of Conscious Experience,* INSERM Symposium no. 6, ed. P. A. Buser and A. Rougeul-Buser, 15–36. Amsterdam: Elsevier/North-Holland.

Kant, Immanuel. 1987 (1790). *Critique of Judgment.* Ed. Werner Pluhar. Indianapolis, IN: Hackett Publishing.

———. 1997 (1788). *Critique of Practical Reason.* Ed. Mary J. Gregor and Andrew Reath. Cambridge: Cambridge University Press.

———. 1999 (1781). *Critique of Pure Reason.* Ed. Paul Guyer and Allen W. Wood. Cambridge: Cambridge University Press.

Kardiner, Abram, and Lionel Ovesey. 1951. *The Mark of Oppression: A Psychological Study of the American Negro.* New York, NY: W. W. Norton & Co.

Khilstrom, J. F. 1987. "The Cognitive Unconscious." *Science,* 137: 1445–51.

Kornhuber, H. H. 1988. "The Human Brain: From Dream and Cognition to Fantasy, Will, Conscience, and Freedom." In *Information Processing by the Brain,* ed. Hans J. Markowitsch, 241–58. Toronto: Hans Huber Publishers.

Kosslyn, S. M. 1980. *Image and Mind.* Cambridge, MA: Harvard University.

———. 1981. "Species of Visual Displays." *Journal of Mental Imagery,* 5:24–26.

———. 1983. *Ghosts in the Mind's Machine.* New York, NY: W. W. Norton.

———. 1994. *Image and Brain: The Resolution of the Imagery Debate.* Cambridge, MA: MIT Press.

Kövecses, Z. 2000. *Metaphor and Emotion: Language, Culture, and Body in Human Feeling.* Cambridge and New York: Cambridge University Press.

———. 2002. *Metaphor: A Practical Introduction.* Oxford: Oxford University Press.

———. 2003. "Language, Figurative Thought, and Cross-cultural Comparison." *Metaphor and Symbol,* 18: 311–20.

———. 2004. "Introduction: Cultural Variation in Metaphor." *European Journal of English Studies,* 8: 263–74.

———. 2005. *Metaphor in Culture: Universality and Variation.* Cambridge: Cambridge University Press.

Krstic, K. 1964. "Marko—The Author of the Term 'Psychology.'" *Acta Instituti Psychologici Universitatis Zagrabiensis,* 36: 7–13.

Kuijsten, Marcel, ed. 2007. *Reflections on the Dawn of Consciousness: Julian Jaynes's Bicameral Mind Theory Revisited.* Henderson, NV: Julian Jaynes Society.

Külpe, Oswlad. 1893. *Grundriss Der Psychologie auf experimenteller Grundlage dargestellt.* Leipzig: Wilhelm Engelmann.

———. 1895. *Einleitung in Die Philosophie.* Leipzig: S. Hirzel.

———. 1920. *Vorlesungen über Psychologie.* Leipzig: S. Hirzel.

Küpfmüller, Karl. 1962. "Nachrichtenverarbeitung im Menschen." In *Taschenbuch der Nachrichtenverabeitung,* ed. K. Steinbuch, 1481–1501. Berlin: Springer.

———. 1971. "Grundlagen der Informationstheorie und Kybernetik." In *Physiologie des Menschen,* vol 10, ed. O. H. Grauer, K. Kramer, and R. Jung, 209–48. Munich: Urban & Schwarzenberg.

Lakoff, George, and Mark Johnson. 1980. *Metaphors We Live By*. Chicago: University of Chicago Press.

Lapointe, F. H. 1970. "Origin and Evolution of the Term 'Psychology.'" *American Psychologist*, 25: 640–46.

Lasch, Christopher. 1979. *The Culture of Narcissism: American Life in an Age of Diminishing Expectations*. New York, NY: W. W. Norton & Co.

Leary, David E., ed. 1990a. *Metaphors in the History of Psychology*. Cambridge: Cambridge University Press.

———. 1990b. "Psyche's Muse: The Role of Metaphor in the History of Psychology." In *Metaphors in the History of Psychology*, ed. David E. Leary, 1–78. Cambridge: Cambridge University Press.

LeBarge, D., and S. J. Samuels. 1974. "Toward a Theory of Automatic Information Processing in Reading." *Cognitive Psychology*, 6:293–323.

LeDoux, Joseph E. 1985. "Brain, Mind and Language." In *Brain and Mind*, ed. David A. Oakley, 197–216. London: Methuen.

Lehrl, S., and B. Fischer. 1985. "De maximale zentrale Informationsfluss beiKüpfmüller und Frank: beträgt er 0 bit/s oder 167 bit/s?" *Grundlagenstudien asu Kybernetik und Geistenswissenschaft/Humankybernetik*, 16:147–54.

Libet, Benjamin. 1965. "Cortical Activation in Conscious and Unconscious Experience." *Perspectives in Biology and Medicine*, 9:77–86.

———. 1981. "The Experimental Evidence for Subjective Referral of a Sensory Experience Backwards in Time: Reply to P. S. Churchland." *Philosophy of Science*, 48:182–97.

———. 1985a. "Unconscious Cerebral Initiative and the Role of Conscious Will in Voluntary Action." *The Behavioral and Brain Sciences*, 8:529–66.

———. 1985b. "Subjective Antedating of a Sensory Experience and Mind-Brain Theories: Reply to Honderich." *Journal of Theoretical Biology*, 114:53–70.

———. 1989. "Conscious Subjective Experience vs. Unconscious Mental Functions: A Theory of the Cerebral Processes Involved." In *Models of Brain Function*, ed. R. M. J. Cotterill, 35–49. Cambridge: Cambridge University Press.

———. 1990. "Time-Delays in Conscious Processes." *The Behavioral and Brain Sciences*, 13:672.

———. 1991. "Conscious vs. Neural Time." *Nature*, 352:27.

———. 1992. "Models of Conscious Timing and the Experimental Evidence." *The Behavioral and Brain Sciences*, 15:213–75.

Libet, Benjamin, W. W. Alberts, E. W. Wright, Jr., B. Feinstein. 1967. "Responses of Human Somatosensory Cortex to Stimuli below Threshold for Conscious Sensation." *Science*, 158:1597–1600.

———. 1972. "Cortical and Thalamic Activation in Conscious Sensory Experience." In *Neurophysiology Studied in Man*, ed. G. G. Somjen, 157–68. Amsterdam: Excerpta Medica.

Libet, Benjamin, L. D. Delatytre, and G. Levin. 1964. "Production of Threshold Levels of Conscious Sensation by Electrical Stimulation of Human Somatosensory Cortex." *Journal of Neurophysiology*, 27: 546–78.

Libet, Benjamin, Curtis A. Gleason, Elwood W. Wright, and Dennis K. Pearl. 1983. "Time of Conscious Intention to Act in Relation to Onset of Cerebral Activity (Readiness-potential)." *Brain*, 106:623–42.

Libet, Benjamin, Dennis K. Pearl, David Morledge, Curtis A. Gleason, Yoshio Morledge, and Nicholas M. Barbaro. 1991. "Control of the Transition from Sensory Detection to Sensory Awareness in Man by the Duration of a Thalamic Stimulus." *Brain*, 14:1731–57.

Libet, Benjamin, Elwood W. Wright Jr., Bertram Feinstein, and Dennis Pearl. 1979. "Subjective Referral of the Timing for a Conscious Sensory Experience." *Brain*, 102: 193–224.

Locke, John. 1975 (1690). *An Essay Concerning Human Understanding*. Oxford: Clarendon Press.

———. 1980 (1690). *Second Treatise of Government*. Ed. C. B. Macpherson. Cambridge: Hackett Publishing.

———. 1983 (1689). *A Letter Concerning Toleration*. Ed. James Tully. Cambridge: Hackett Publishing.

———. 1988 (1690). *Two Treatises of Government*. Ed. P. Laslett. Cambridge: Cambridge University Press.

Lutz, Catherine A. 1988. *Unnatural Emotions: Everyday Sentiments on a Micronesian Atoll and Their Challenge to Western Theory*. Chicago: University of Chicago Press.

MacCormac, Earl R. 1976. *Metaphor and Myth in Science and Religion*. Durham, NC: Duke University Press.

———. 1985. *A Cognitive Theory of Metaphor*. Cambridge, MA: MIT Press.

Maine, Sir Henry Sumner. 1888 (1861). *Ancient Law*. New York: Henry Holt and Company.

Martin L. Gross. 1978. *The Psychological Society: A Critical Analysis of Psychiatry, Psychotherapy, Psychoanalysis and the Psychological Revolution*. New York: Simon and Schuster.

Marx, Karl. 1954 (1848). *Communist Manifesto* (Introduction by Stefan T. Possony). Chicago: Henry Regency Company.

———. 1975 (1867). *Capital: A Critique of Political Economy*. New York: International Publishers.

Matthews, Gareth. 2000. "Internalist Reasoning in Augustine for Mind–Body Dualism." In *Psyche and Soma*, ed. John P. Wright and Paul Potter, 133–46. Oxford: Clarendon Press.

McGee, Micki. 2005. *Self-Help, Inc: Makeover Culture in American Life*. Oxford: Oxford University Press.

McLennan, John F. 1865. *Primitive Marriage: An Inquiry into the Origin of the Form of Capture in Marriage Ceremonies*. Edinburgh: Adam and Charles Black.

McVeigh, Brian J. 1996. "Standing Stomachs, Clamoring Chests and Cooling Livers: Metaphors in the Psychological Lexicon of Japanese." *Journal of Pragmatics*, 26: 25–50.

———. 2006a. "The Self as Interiorized Social Relations: A Jaynesian Approach to Problems of Agency and Volition." In *Reflections on the Dawn of Consciousness: Julian Jaynes's Bicameral Mind Theory Revisited*, ed. Marcel Kuijsten, 203–32. Henderson, NV: Julian Jaynes Society.

———. 2006b. Review of *Self-Help, Inc: Makeover Culture in American Life* (Micki McGee). Oxford: Oxford University Press, 2005. Metapsychology Online Reviews.

———. 2008. Review of *Saving the Modern Soul: Therapy, Emotions, and the Culture of Self-Help* (Eva Illouz). Berkeley: University of California Press, 2008. Metapsychology Online Reviews.

———. 2009. Review of *In Therapy We Trust: America's Obsession with Self-Fulfillment* (Eva S. Moskowitz). Baltimore: Johns Hopkins University, 2001. Metapsychology Online Reviews.

———. 2013. "Mental Imagery and Hallucinations as Adaptive Behavior: Divine Voices and Visions as Neuropsychological Vestiges." *The International Journal of the Image*, 3(1): 25–36.

Mead, G. H. 1967 (1934). *Mind, Self, and Society*. Chicago: University of Chicago Press.

Messinger, Sheldon E., Harold Sampson, and Robert D. Towne. 1990. "Life as Theater: Some Notes on the Dramaturgic Approach to Social Reality." In *Life as Theater: A Dramaturgical Sourcebook*, ed. Dennis Brissett and Charles Edgley, 73–84. New York: Aldine de Gruyter.

Meyer, John W. 1986. "The Self and the Life Course: Institutionalization and Its Effects." In *Human Development and the Life Course*, ed. A. Sorensen et al., 199–216. Hillsdale, NJ: Erlbaum.

———. 2004. "World Society, the Welfare State, and the Life Course: An Institutional Perspective." *SocialWorld—Working Paper No. 8*. Available at http://www.uni-bielefeld.de/soz/Forschung/Projekte/socialworld/pdf/working%20paper%208.pdf.

Michael, Emily. 2000. "Renaissance Theories of Body, Soul, and Mind." In *Psyche and Soma*, ed. John P. Wright and Paul Potter, 147–72. Oxford: Clarendon Press.

Mill, J. S. 1984 (1859). *On Liberty*. Harmondsworth: Penguin.

Mitchell, W. J. T. 1995. *Picture Theory: Essays on Verbal and Visual Representation*. Chicago: University of Chicago Press.

Morgan, Lewis Henry. 1963 (1877). *Ancient Society: Researches in the Lines of Human Progress through Savagery through Barbarism to Civilization*. New York: World Publishing Company.

Morris, Colin. 1972. *The Discovery of the Individual, 1050–1200*. New York: Harper & Row.

Moskowitz, Eva S. 2001. *In Therapy We Trust: America's Obsession with Self-Fulfillment.* Baltimore: Johns Hopkins University.
Müller, Johannes Peter. 1833–40. *Handbuch der Physiologie des Menschen* (2 vols.). Coblenz: Hölscher.
Müller, M. 1867. *Metaphor. Lectures on the Science of Language* 2, 351–99. New York: Scribner.
Mumford, Lewis. 1934. *Technics and Civilization.* Orlando, FL: Harcourt Brace and Company.
Nelson, Victoria. 2001. *The Secret Life of Puppets.* Cambridge, MA: Harvard University Press.
Nisbet, Robert. 1980. *History of the Idea of Progress.* New York: Basic Books.
Nolan, James L. 1998. *The Therapeutic State: Justifying Government at Century's End.* New York: University Press.
Nørretranders, Tor. 1998. *The User Illusion: Cutting Consciousness Down to Size.* New York: Penguin Books.
Oakely, David A., and Lesley C. Eames. 1985. "The Plurality of Consciousness." In *Brain and Mind,* ed. David A. Oakley, 215–51. London: Methuen.
Ong, Walter J. 1982. *Orality and Literacy: The Technologizing of the Word.* New York, NY: Methuen.
Ortony, A., ed. 1979. *Metaphor and Thought.* Cambridge: University of Cambridge Press.
Ozick, Cynthia. 1986. "The Moral Necessity of Metaphor: Rooting History in a Figure of Speech." *Harper's,* 272: 62–68.
Palm, Franklin C. 1936. *The Middle Class Then and Now.* New York: Macmillan.
Pegis, Anton C. 1948. *Introduction to St. Thomas Aquinas: The Essence of the Summa Theologica and the Summa contra Gentiles.* New York, NY: Random House.
Polsky, Andrew J. 1991. *The Rise of the Therapeutic State.* Princeton, NJ: Princeton University Press.
Posner, M. I., and C. R. R. Snyder. 1975. "Attention and Cognitive Control." In *Information Processing and Cognition: The Loyola Symposium,* ed. R. L. Solso, 55–85. Hillsdale, NJ: Lawrence Erlbaum Associates.
Pylyshyn, Z. W. 1973. "What the Mind's Eye Tells the Mind's Brain: A Critique of Mental Imagery." *Psychological Bulletin:* 80, 1–24.
———. 1978. "Imagery and Artificial Intelligence." In *Perception and Cognition: Issues in the Foundations of Psychology* (vol. 9), ed. C. W. Savage, 19–56. Minneapolis: University of Minnesota Press.
———. 1981. "The Imagery Debate: Analogue Media versus Tacit Knowledge." *Psychological Review,* 88: 16–45.
Radding, Charles M. 1985. *A World Made by Men: Cognition and Society, 400–1200.* Chapel Hill: University of North Carolina Press.
Rappaport, Roy A. 1968. *Pigs for the Ancestors: Ritual in the Ecology of a New Guinea People.* New Haven: Yale University Press.
Reed, Edward S. 1997. *From Soul to Mind: The Emergence of Psychology, from Erasmus Darwin to William James.* New Haven, CT: Yale University Press.
Rey, Roselyne. 2000. "Psyche, Soma, and the Vitalist Philosophy of Medicine." In *Psyche and Soma,* ed. John P. Wright and Paul Potter, 255–66. Oxford: Clarendon Press.
Richards, Graham. 1992. *Mental Machinery, Part 1: The Origins and Consequences of Psychological Ideas from 1600 to 1850.* London: Athlone.
Rieff, Philip. 1966. *The Triumph of the Therapeutic.* New York, NY: Harper & Row.
Rifkin, Jeremy. 2004. *The European Dream: How Europe's Vision of the Future is Quietly Eclipsing the American Dream.* New York: Tarcher/Penguin.
Robinson, Daniel N. 1995. *An Intellectual History of Psychology.* Madison, WI: University of Wisconsin Press.
Robinson, T. M. 2000. "The Defining Features of Mind–Body Dualism in the Writings of Plato." In *Psyche and Soma,* ed. John P. Wright and Paul Potter, 37–56. Oxford: Clarendon Press.
Rose, N. 1996. *Inventing Our Selves. Psychology, Power and Personhood.* Cambridge: Cambridge University Press.

Rosenberg, Daniel, and Susan Harding, eds. 2005. *Histories of the Future.* Durham, NC: Duke University Press.
Rosenthal, Peggy. 1984. *Words and Values: Some Leading Words and Where They Lead Us.* New York: Oxford Press.
Ryle, Gilbert. 1949. *The Concept of Mind.* New York: Hutchinson's University Library.
Sahlins, Marshall. 1976. *Culture and Practical Reason.* Chicago: University of Chicago Press.
Sahlins, Marshall D., and Elman R. Service, eds. 1969. *Evolution and Culture.* Ann Arbor: University of Michigan Press.
Saint-Simon, Claude Henri de. 1964 (1825). "New Christianity." In *Social Organization: The Science of Man and Other Writings*, ed. and trans, Felix Markham, 81–116. New York: Harper Torchbooks.
Schneider, W., and R. M. Shiffrin. 1977. "Controlled and Automatic Human Information Processing: I. Detection, Search, and Attention." *Psychological Review,* 84:1–66.
Shibles, W. A. 1971. *Metaphor: An Annotated Bibliography and History.* Whitewater, WI: Language Press.
Shiffrin, R. M., and W. Schneider. 1977. "Controlled and Automatic Human Information Processing: II. Perceptual Learning, Automatic Attending, and a General Theory." *Psychological Review,* 84:127–90.
Shilling, C. 1994. *The Body and Social Theory.* London: Sage Publications.
Shweder, Richard A. 1990. "Cultural Psychology: What Is It?" In *Cultural Psychology: Essays on Comparative Human Development,*" ed. James W. Stigler, Richard A. Shweder, and Gilbert Herdt, 1–43. Cambridge: Cambridge University Press.
Smith, M. B. 1985. "The Metaphorical Basis of Selfhood." In *Culture and Self: Asian and Western Perspectives,* ed. A. J. Marsella, G. DeVos, and F. L. K. Hsu, 6–88. London: Tavistock.
Solomons, L. M., and G. Stein. 1886. "Normal Motor Automatism." *Psychological Review,* 3:492–512.
Soskice, J. M. 1985. *Metaphor and Religious Language.* Oxford: Clarendon Press.
Spencer, Herbert. 1862. *Synthetic Philosophy: First Principles.* New York: De Witt Revolving Fund.
———. 1864–67. *The Principles of Biology* (2 vols.). New York: D. Appleton.
———. 1870. *The Principles of Psychology* (2 vols.). New York: D. Appleton.
———. 1876–96. *The Principles of Sociology* (3 vols.). New York: D. Appleton.
———. 1892–93. *The Principles of Ethics* (2 vols.). New York: D. Appleton.
Spengler, Oswald. 1945 (1926–28). *Decline of the West.* New York: Knopf.
Steward, Julian H. 1936. *Economic and Social Basis of Primitive Bands.* Berkeley: University of California Press.
———. 1953. "Evolution and Process." In *Anthropology Today,* ed. Alfred L. Kroeber, 313–26. Chicago: University of Chicago Press.
———. 1955. *Theory of Culture Change.* Urbana: University Press.
———. 1977. *Evolution and Ecology.* Urbana: University of Illinois Press.
Szamosi, Géza. 1986. *The Twin Dimensions: Inventing Time and Space.* New York, NY: McGraw Hill Book Co.
Terman, L. M. 1938. *Psychological Factors in Marital Happiness.* New York, NY: McGraw-Hill.
Terman, L. M., P. Buttenwieser, L. W. Ferguson, W. B. Johnson, and D. P. Wilson. 1938. *Psychological Factors in Marital Happiness.* New York and London: McGraw-Hill Book Company.
Tester, Keith. 1992. *Civil Society.* London: Routledge.
Therborn, Goran. 1995. *European Modernity and Beyond: The Trajectory of European Societies, 1945–2000.* London: Sage Publications.
Trickner, Dietrich. 1966. "Aufnahme, Speicherung und Verabeitung von Information durch den Menschen." *Veröffentlichungen der Schleswig-holsteinischen Universitätsgesellschaft, Neue Folge,* no. 44. Kiel: Verlag Ferdinand Hirt.
Trilling, Lionel. 1972. *Sincerity and Authenticity.* Cambridge, MA: Harvard University Press.
Tucker, James. 1999. *The Therapeutic Corporation.* New York: Oxford University Press.

Turgot, Anne Robert Jacques. 1844 (1750). *Plan de Deux Discours sur l'Histoire Universelle*. Paris: Guillaumin.

Turner, B. S. 1983. *Religion and Social Theory*. London: Heinemann Educational Books.

Turner, Victor. 1967. *The Forest of Symbols: Aspects of Ndembu Ritual*. Ithaca, NY: Cornell University Press.

Twenge, Jean, and W. Keith Campbell. 2009. *The Narcissism Epidemic: Living in the Age of Entitlement*. New York, NY: Free Press.

Tylor, Edward Burnett. 1958 (1871). *Primitive Culture: Researches into the Development of Mythology, Philosophy, Religion, Languages, Art and Customs*. New York: Harper Torchbooks.

Van der Eijk, Philip J. 2000. "Aristotle's Psycho-physiological Account of the Soul–Body Relationship." In *Psyche and Soma*, ed. John P. Wright and Paul Potter, 57–78. Oxford: Clarendon Press.

Van Noppen, J. P., S. De Knop, and R. Jogen. 1985. *Metaphor: A Bibliography of Post-1970 Publications*. Amsterdam: Benjamins.

Vera, A. H., and H. A. Simon. 1993. "Situated Action: A Symbolic Interpretation." *Cognitive Science*, 17:7–48.

Vico, Giambattista. 1961 (1725). *The New Science of Giambattista Vico*. Translated from 3rd edition of 1744, Thomas Goddard Bergin and Max Harold Fisch. New York: Doubleday.

Vidal, Fernando. 1993. "Psychology in the 18th Century: A View from Encyclopaedias." *History of the Human Sciences*, 6(1): 89–119.

Von Staden, Heinrich. 2000. "Body, Soul, and Nerves: Epicurus, Herophilus, Erasistratus, the Stoics, and Galen." In *Psyche and Soma*, ed. John P. Wright and Paul Potter, 79–116. Oxford: Clarendon Press.

Voss, Stephen. 2000. "Descartes: Heart and Soul." In *Psyche and Soma*, ed. John P. Wright and Paul Potter, 173–96. Oxford: Clarendon Press.

Ward, Steven C. 2002. *Modernizing the Mind. Psychological Knowledge and the Remaking of Society*. Westport, CT: Praeger.

Warner, Rex. 1963. *The Confessions of St. Augustine*. Translated by Rex Warner. New York: Mentor.

Weber, Max. 1978 (1922). *Economy and Society: An Outline of Interpretive Sociology* (2 vols.). Berkeley: University of California Press.

Weintraub, Karl Joachim. 1978. *The Value of the Individual: Self and Circumstance in Autobiography*. Chicago: University of Chicago Press.

Werner, Wolf, ed. 2011. *Metareferential Turn in Contemporary Arts and Media: Forms, Functions, Attempts at Explanation* (in collaboration with Katharina Bantleon and Jeff Thoses). Amsterdam/New York: Rodopi.

Wertheim, Margaret. 1999. *Pearly Gates of Cyberspace*. New York: W. W. Norton.

White, Leslie. 1959. *The Evolution of Culture*. New York: Farrar Straus.

Whitten, Andrew, ed. 1991. *Natural Theories of Mind: Evolution, Development and Simulation of Everyday Mindreading*. Cambridge, MA: Blackwell.

Wilber, Ken. 2007. *Up from Eden: A Transpersonal View of Human Evolution*. Wheaton, IL: Quest Books.

Williams, Raymond. 1983. *Keywords: A Vocabulary of Culture and Society*. New York: Oxford University Press.

Woodward, William, and June Tower. 2007. "Julian Jaynes: Introducing His Life and Thought." In *Reflections on the Dawn of Consciousness: Julian Jaynes's Bicameral Mind Theory Revisited*, Marcel Kuijsten, 13–74. Henderson, NV: Julian Jaynes Society.

Wright, John P. 2000. "Substance versus Function Dualism in Eighteenth-Century Medicine." In *Psyche and Soma*, ed. John P. Wright and Paul Potter, 237–254. Oxford: Clarendon Press.

Wright, John P., and Paul Potter. 2000. "Introduction." In *Psyche and Soma*, ed. John P. Wright and Paul Potter, 1–11. Oxford: Clarendon Press.

Wundt, Wilhelm M. 1900–1920. *Völkerpsychologie: eine Untersuchung der Entwicklungsgesetze von Sprache, Mythus und Sitte* (*Ethno-Cultural Psychology: An Investigation of the Developmental Laws of Language, Myth, and Customs*). Leipzig: Kröner.

———. 1893 (1904). *Principles of Physiological Psychology*. Translated by E. B. Titchener. London: Allen.
———. 1916. *Elements of Folk-Psychology*. Translated by E. L. Schaub. London: Allen.
———. 1969 (1897). *Outlines of Psychology*. Trans C. H. Judd. St. Clair Shores, MI: Scholarly Press.
———. 1973 (1912). *An Introduction to Psychology*. New York: Arno Press.
Wyer, R. S., Jr., ed. 1997. *The Automaticity of Everyday Life.* Mahwah, NJ: Lawrence Erlbaum Associates.
Yates, Frances A. 1966. *The Art of Memory.* Chicago, IL: University of Chicago Press.
Zerubavel, Eviatar. 1981. *Hidden Rhythms.* Chicago, IL: University of Chicago Press.
Zimmerman, Manfred. 1986. "Neurophysiology of Sensory Systems." In *Fundamentals of Sensory Physiology,* ed. Robert F. Schmidt, 68–116. Berlin: Springer-Verlag.
———. 1989. "The Nervous System in the Context of Information Theory." In *Human Physiology* (2nd ed.), ed. R. F. Schmidt and G. Thews, 166–73. Berlin: Springer-Verlag.

Index

Abélard, Peter, 148
abstracting, viii, 4, 11, 12, 15, 19, 20, 21, 22, 24, 42, 44, 45, 53, 59, 62, 64, 67, 89, 90, 92, 100, 111, 112, 114, 128, 129, 130, 146, 149, 161, 163, 169, 195, 197
Ach, Narziss, 171
Adorno, Theodore, 180
Alberti, Leon Battista, 63
Alhazen (Ibn al-Haytham), 66
Al-Kindi (Abu-Yusuf Ya'qub ibn-Isha), 66
Alsted, Johann Heinrich, 72
American Revolutionary Way, 128
analogizing, 4, 22, 31, 35, 36, 41, 61, 66, 95n8, 113, 158, 161, 162, 167
Andersen, Hans Christian, 137
anima, 45, 51, 64, 71, 76n3, 79, 81, 149. *See also* animus
animal gravitation, 98. *See also* animal magnetism
animal magnetism, 98. *See also* animal gravitation
animal spirits, 66, 80, 92
animism, 32, 64, 80, 88, 132
animus, 71, 76n3. *See also* anima
Anselm, Archbishop of Canterbury, 148
Aquinas, St. Thomas, 53, 57n15
Arabic world, 59, 114
Aristotle, 41
associationism, 23, 51, 87, 90, 92, 93, 95n7, 102
associationists, 51
astroscopic, 55
Atlantis, 60, 120, 121
Augustus (Emperor), 157
Aurelius, Marcus, 157
authenticity, 114, 177, 186
automatic writing, 98

Bacon, Sir Francis, 90, 120, 121
Bain, Alexander, 173
Barfield, Owen, 19
behaviorism, vii, 41, 76, 83, 196n12
Bell, Sir Charles, 73, 101
Bellamy, Edward, 138
Berkeley, George, 73, 92
Bessel, Fredrich, 99
Bichat, M. F. X., 51
Biondo, Flavio, 114
Biran, François-Pierre-Gonthier Maine de, 85
Blanchard, Phyllis, 180
Bodin, Jean, 119
body, 10, 12, 15, 25, 29n12, 32, 34, 43, 44, 45, 46, 47n10, 51, 52, 53, 54, 55, 56, 66, 71, 72, 73, 74, 75, 77, 78, 79, 80, 81, 82, 83, 84, 87, 88, 89, 90, 91, 93, 94, 97, 98, 100, 101, 102, 103, 104, 129, 143, 169, 173, 194, 195, 202, 206
Bois-Reymond, Emil du, 100, 173
Boorstin, Daniel, 62, 67n1, 67n4, 68n6, 68n13, 68n21, 68n22, 68n23, 115n1,

115n2, 115n4, 115n10, 115n15, 115n16, 115n17, 124n5
Boyle, Robert, 90
Bradley, James, 99
Brentano, Franz, 73, 172, 173
British empiricism, 87
Broca, Paul, 75
Brunelleschi, Filippo, 63
Buddhism, 55, 56, 111
Buffon, Georges Louis Leclerc Comte de, 121
Burton, Robert, 74
Bury, J. B., 17n12

Cabanis, Pierre Jean Georges, 83
calendars, 110, 112, 113, 130
Campbell, W. Keith, 190
capitalism, 16, 113, 123, 151, 189, 190, 195
Cartesian dualism, 88, 143
cartographers, 60, 64
Casmann, Otto, 72
Charcot, Jean Martin, 98
Childe, Gordon V., 132
China, 61, 109, 113, 124
Chinese civilization, 54; cosmographers of, 61; history of, 109; Psychology and, 55; response to interiorization, 53
Christian theology, 52
Christianity, 45, 53, 111, 129, 130, 145
Cicero, 50
clocks, 99, 113
Comte, Auguste, 121, 128, 129, 130, 131
conceptualism, 148
conciliation, 35
conscious interiority, 10, 16, 26, 31, 32, 33, 34, 49, 76, 98, 100, 101, 104, 105, 112, 143, 144, 167, 170, 172, 177, 179, 183, 199, 201, 202, 203, 204, 205, 207. *See also* consciousness; interiority; interiorization
Condorcet, Marie Jean de, 121
Confucianism, 55, 109, 111, 124
consciousness, vii, viii, 4, 6, 16, 19, 25, 31, 32, 33, 35, 49, 54, 56, 78, 84, 94, 101, 122, 133, 145, 146, 158, 165, 170, 172, 174, 195, 201, 202, 203, 204, 205, 206, 207, 208n3. *See also* conscious interiority; interiority; interiorization
constitutionalism, 118, 123
consumerism, 139, 179, 190, 195
Cook, James, 60
Copernicus, Nicolaus, 21, 64
cosmic rupture, 78, 79, 87, 98. *See also* cosmic split
cosmic split, 9, 45, 54, 55, 77, 78, 79, 87, 92. *See also* cosmic rupture
cosmic worldview, 21. *See also* cosmology; cosmos
cosmology, 9, 21, 24, 27, 43, 54, 57n16, 59, 63, 64, 66, 74, 77, 112, 156, 166, 195, 197. *See also* cosmic worldview, cosmos
cosmos, 3, 4, 11, 14, 15, 19, 21, 24, 26, 54, 59, 61, 62, 64, 67, 104, 113, 148. *See also* cosmic worldview; cosmology
cyberspace, 3, 15, 21, 195

d'Oresme, Nicole, 113
Darwin, Charles, vii, 93, 121, 139
democracy, 36, 120, 123, 128, 189
Descartes, René, 73, 79, 80, 83, 90, 92, 121
Diderot, Denis, 120
diffusionism, 132
Dionysus Exiguus, 119
dualism, 5, 12, 23, 34, 43, 44, 45, 46, 52, 53, 55, 71, 77, 78, 79, 80, 82, 83, 84, 87, 88, 89, 91, 93, 94, 97, 98, 99, 101, 104, 128, 143, 147, 169, 170, 172, 193, 194, 196n13; continuous, 44, 78; discontinuous, 53, 55, 77, 78, 79; strong, 77, 78; weak, 77. *See also* mind–body
Durkheim, Émile, 130, 132, 166, 203
dystopia, 138

El Dorado, 60
Elias, Norbert, 6
emancipatory ethos, 123, 124, 127, 128
embourgeoisement, 14
emotion, 20, 23, 24, 27, 35, 43, 44, 47n10, 52, 53, 55, 79, 91, 110, 120, 165, 166, 173, 177, 178, 179, 184, 185, 186, 187, 188, 191n4, 195, 197, 199, 206
empiricism, 55, 87, 90, 91, 92, 94, 95, 161, 172
empiricist-versus-rationalist debate, 88
empsychologia, 72

Empyrean (of God), 61
Enlightenment, 72, 90, 91, 103, 114, 119, 120, 121, 123, 127, 128, 135, 136, 194, 206
Epicurus of Samos, 46, 47n17
Erasistratus of Cyrene, 46, 47n17
Esdaile, James, 98
eternity, 61, 119
ether, 51, 61
Euhemerus, 110
evolution, 131, 132, 133, 139, 152n3, 199; social, 131, 139
excerption, 35, 36, 145
experimental aesthetics, 103

Fechner, Gustav, 103, 104
Fechner–Weber's Law, 104
figurative metaphors, 22, 29n12
Flourens, Pierre, 73, 75
Fontenelle, Bernard le Bovier de, 120
Foucault, Michel, 151, 153n38, 179
Francesca, Piero della, 63
Frazer, James, 132
French Revolution, 118, 121, 127
Freud, Sigmund, 94, 98, 132, 184, 206
Fugger, Jakob, 113
Furedi, Frank, 184
future, 5, 12, 14, 15, 26, 35, 36, 80, 109, 111, 112, 113, 118, 120, 121, 127, 129, 135, 137, 138, 140n3, 140n4, 144, 156, 158, 166, 195

Galen, 28, 47n17, 66, 101
Galileo Galilei, 21, 65, 90, 103, 113
Gall, Franz Joseph, 52
Garden of Eden, 60
geography, 4, 59, 60, 79, 193
German nativism, 87
Gestalt Psychology, 81
Giambattista, Vico, 120
Gilbert, Ryle, 83
glossolalia, 98
Goelenius, Rudolphus (Rudolf Goeckel), 72
Goffman, Irving, 162, 167n5
Gog and Magog, 60
Greeks (ancient), 28, 28n9, 41, 43, 44, 47n10, 52, 54, 63, 66, 72, 74, 94, 110, 119, 120, 121, 135
Gross, Martin L., 177, 184
Groves, Ernest, 180

Harris, Marvin, 132, 133n7
Hartley, David, 92
Harvey, William, 66
Heaven, 3, 21, 54, 61, 62, 64, 82, 92, 112, 183, 195
Hecataeus of Miletus, 110
Hegel, Wilhelm Friedrich, 122, 124n7
Hell, 3, 21, 60, 61, 193, 195
Hellenistic thought, 46, 99
Helmholtz, Hermann von, 73, 94, 100, 101, 102, 170
Herbart, Johann Friedrich, 94, 101
Herodotus, 14, 110, 120
Herophilus, 46, 47n17
Hesiod, 109
Hinduism, 55, 111
history, 56n4, 73, 74, 79, 91, 93, 109, 110, 111, 114, 117, 118, 119, 120, 121, 122, 124n7; objective, 109. *See also* progress
Hobbes, Thomas, 73, 90
Homer, 44, 149
Hong Xiuquan, 124
Hume, David, 73, 91, 92
humors, 27, 28, 56
Huxley, Aldous, 44
hypnosis, 98, 105n3, 162

"I", 35, 36, 42, 52, 64, 94, 105n3, 144, 145, 156, 158, 162, 163, 164, 167, 167n2, 206. *See also* "me"; self
Ibn Khaldun, 109
identity politics, 16, 181
Illouz, Eva, 177, 179, 182, 182n8, 182n9, 182n19, 185, 186, 187, 190, 191n8, 191n14
imageless thought, 42, 171, 172
imagery, 23, 41, 42, 51, 74, 76, 98, 148, 171, 196n14
Indian civilization, 55, 56, 93, 110, 117; response to interiorization, 55
individualism, 78, 128, 143, 149, 151, 153n25, 178, 179, 185, 188
individuality, 43, 45, 149, 150, 152, 155, 156, 157, 195; emergence of, 149

individuation, 36, 83, 90, 97, 145, 151, 152, 153n37, 156
industrialism, 16, 26, 113, 118, 123, 128, 129, 132, 144, 151
intention, 6, 36, 55, 56, 90, 144, 145, 146, 147, 148, 149, 162, 172, 173, 177, 185, 199, 203
interiority, 4, 5, 10, 26, 31, 32, 37n3, 41, 42, 43, 44, 47n10, 49, 50, 51, 52, 53, 54, 55, 56, 63, 64, 66, 74, 82, 83, 84, 85, 87, 88, 89, 90, 91, 92, 94, 95, 97, 99, 101, 102, 104, 105n3, 112, 117, 120, 122, 123, 133, 136, 143, 144, 145, 146, 147, 148, 149, 150, 151, 152, 156, 157, 158, 161, 163, 164, 165, 166, 167n6, 167n7, 194, 195, 196n9, 199; construction of self, 16; definition of, 31, 33, 201, 202, 203, 204, 205, 206, 207; emergence of proto-Psychology, 74, 75, 76; features of, 34, 35, 36; modern Psychology, 79, 80, 169, 170, 171, 172, 173, 174; psychotherapy, 183, 184, 190; roles, 166, 167; social adaptation, 33, 34; therapeutic society, 177, 178, 179. *See also* conscious interiority; consciousness; interiorization
interiorization, 4, 5, 16, 33, 34, 51, 52, 53, 55, 63, 74, 75, 79, 82, 83, 94, 123, 145, 151, 161, 166, 169, 172, 173, 184, 190, 194, 206. *See also* conscious interiority; consciousness; interiority; spatialization, of mind
introception, 10, 17n4, 35, 41, 42, 64, 91, 97, 163, 196n14
introcosm, 9, 10, 11, 12, 27, 34, 35, 42, 51, 63, 64, 71, 74, 87, 91, 92, 97, 148, 156, 162, 193, 194, 195. *See also* macrocosm; microcosm
introscape, 35, 64, 194
introscopic, 11, 12, 59, 74, 97, 98, 194, 195. *See also* macroscopic; microscopic; scopic
introspection, 11, 17n4, 25, 36, 41, 42, 46, 47n10, 51, 75, 76, 79, 85, 92, 97, 145, 158, 170, 172, 201
introspectionists, 51
Islamic world, 61, 109, 111

Jainism, 55
James I, King of Aragon, 156
James, William, 170, 208n13
Jaynes, Julian, vii, viii, ix, 3, 6, 7n2, 24, 28n10, 29n15, 29n16, 29n17, 29n19, 31, 32, 34, 36n1, 37n4, 37n5, 37n6, 37n7, 37n9, 37n10, 41, 47n3, 47n9, 47n10, 56n1, 56n3, 133, 156, 159n6, 163, 167n4, 167n6, 177, 182n17, 190, 207n1, 208n3, 208n17, 208n25, 208n30, 208n31, 208n32
Juan Luis Vives, 51, 119
Judeo–Christian eschatology, 127
judgments of God, 147

Kant, Immanuel, 94, 95, 95n11, 101, 165, 206
Kardiner, Abram, 181
Kepler, Johannes, 21, 64, 90
Koffka, Kurt, 173
Kuijsten, Marcel, viii
Külpe, Oswald, 133n5, 171

Lasch, Christopher, 190
Le Rond d'Alembert, Jean, 120
Leibniz, Gottfried, 72, 73
Lesseps, Ferdinand de, 129
liberalism, 123
literal metaphors, 121
Locke, John, 11, 73, 79, 91, 92, 165; as proto-Psychologist, 90
Lotze, Rudolph Hermann, 102
Lough Derg, 61
Lu Xiangshan, 55
Lytton, Bulwer, 138

macrocosm, 9, 10, 12, 21, 34, 51, 61, 62, 92, 194, 195. *See also* introcosm; microcosm
macroscopic, 11, 12, 21, 59, 62, 63, 64, 65. *See also* introscopic; microscopic; scopic
Maine, Henry James Sumner, 131
Malthus, Thomas, 138
Mao Zedong, 124
maps, 59, 60, 61, 64, 194
Marbe, Karl, 172
Marcus Fabius Quintilianus, 50
Marden, Orison Swett, 144

Marulic, Marko, 72
Marx, Karl, 132
Masaccio, Tommaso, 63
materialism, 89, 100, 103, 169
McGee, Micki, 177, 178, 180, 182n5, 182n10, 187, 188, 189, 190, 191n3, 191n11, 191n12, 191n13, 191n15, 191n16
McLennan, John Ferguson, 131
"me", 35, 36, 42, 52, 64, 94, 105n3, 144, 145, 156, 158, 162, 163, 164, 167, 167n2, 206. *See also* "I"; self
measurement, 10, 11, 59, 62, 63, 99, 100, 112, 161, 189
meta-framing and, 67; of psyche, 97
Mechthild of Magdeburg, 193
Medieval cosmology, 64; collapse of, 63
Melanchton, Philipp, 72
mental images, 41, 90, 102. *See also* mental representations
mental photography, 178
mental representations, 41, 56. *See also* mental imagery
mentalité, 6, 10; Medieval, 61
Mercator, Gerardus, 59, 60
Mercier, Sebastien, 137
Mesmer, Friedrich Anton, 98
mesmerism, 73, 98
meta-framing, 4, 19, 20, 21, 22, 31, 42, 45, 52, 53, 59, 60, 62, 64, 67, 74, 79, 82, 88, 90, 97, 118, 127, 128, 135, 146, 158, 161, 162, 163, 164, 165, 170, 195, 206; history and, 109, 111, 112; measurement, 67; metaphors and, 26, 28. *See also* reframing
meta-language, 20
meta-narratives, 6
metaphier, 22, 24, 26, 27, 44, 49, 66, 78, 79, 186, 187
metaphor, vii, viii, 3, 4, 10, 13, 14, 16, 17n3, 19, 20, 21, 22, 23, 24, 25, 26, 27, 28, 28n9, 28n11, 29n12, 29n13, 29n14, 29n20, 41, 44, 47n10, 49, 55, 56n6, 74, 77, 83, 112, 113, 114, 129, 130, 136, 162, 171, 186, 195, 206; and interiority, 34, 36, 102, 105n3; and meta-framing, 23, 26; of mind, 11, 12, 13, 31, 32; and mind–words, 23, 24, 79, 84, 99. *See also* metaphier; metaphrand; paraphier; paraphrand
metaphrand, 22, 24, 26, 27, 29n15, 44, 49, 66, 78, 79, 186, 187
microcosm, 9, 10, 12, 21, 34, 51, 62, 162, 195. *See also* introcosm; macrocosm
macroscopic, 11, 12, 21, 59, 62, 63, 64, 65. *See also* introscopic; macroscopoic; scopic
Middle Ages, 21, 61, 71, 114, 145, 147, 148, 155, 157, 158, 166, 193
Mill, James, 73, 93
Mill, John Stuart, 73, 93
millenarianism, 15, 129, 135, 139
mind, 17n10, 22, 23, 25, 26, 31, 32, 34, 36, 41, 42, 43, 45, 49, 51, 52, 54, 55, 56, 56n4, 60, 61, 62, 63, 64, 71, 72, 73, 74, 75, 76n3, 78, 79, 80, 81, 82, 83, 84, 87, 88, 89, 90, 91, 92, 93, 94, 95, 97, 98, 99, 100, 101, 102, 103, 104, 111, 112, 121, 122, 132, 133, 137, 138, 147, 148, 149, 152n3, 156, 163, 169, 171, 172, 173, 178, 179, 184, 191n4, 194, 195, 196n12, 197, 198, 199, 203, 205, 206, 208n13; as central processing mechanism (CPM), 11, 197, 198, 199, 206; as container (metaphor), 11, 13; and historical forces, 11; as mirror (metaphor), 13; its plasticity, 12, 13, 146, 198; as putty (metaphor), 11, 12, 13, 199; and social pragmatics, 13, 198; as tool (metaphor), 13; as vase (metaphor), 11, 12, 13, 199. *See also* psyche
mind cure, 178
mind's eye, 10, 17n3, 17n4, 41, 42, 64, 151, 178
mind–body dualism, 46, 47n13, 52, 53, 55, 56n13, 77, 78, 80, 81, 82, 83, 88, 93, 94, 99, 101, 102, 173
mind–body interaction, 80, 82, 92
mind–space, 105n3, 145, 151, 152, 158, 162, 171
modernity, 5, 15, 35, 59, 64, 117, 118, 123, 124, 127, 151, 166, 178, 179, 180, 185, 194, 195
monism, 44, 46, 77, 139
Morgan, J. P., 113
Morgan, Lewis Henry, 131

Moskowitz, Eva, 177, 178, 179, 180, 181, 182n2, 182n3, 182n4, 182n6, 182n7, 182n11, 182n12, 182n13, 182n14, 182n15, 182n16, 182n18, 185, 190, 191n1, 191n6, 191n7
Muhammad, 109
Müller, Johannes P., 73, 102, 170
Mumford, Lewis, 67n3, 68n5, 68n8, 68n9, 68n17, 68n19

nanoscopic, 15, 21. *See also* scopic
narcissism, 185, 190
narratization, 35, 36, 144, 156, 158
national statism, 123, 139
Nelson, Victoria, 193, 194, 196n6, 196n7, 196n8, 196n9
Neo-Luddites, 139
Neoplatonism, 45, 147
Newton, Sir Isaac, 11, 21, 65, 75, 91, 92, 94, 98, 101, 113, 120
Nietzsche, Friedrich, 26
Nolan, James L., 184
nominalism, 148
nonconscious, 33, 35, 42, 56, 87, 94, 99, 100, 101, 167n2, 174, 202, 203, 204, 206, 207. *See also* unconscious
non-Euclidean geometry, 15
Nørretranders, Tor, 37n3, 37n8, 56n2, 167n2, 204, 207n2, 208n4, 208n6, 208n7, 208n9, 208n10, 208n18, 208n19, 208n20, 208n23, 208n24, 208n27, 208n34, 208n35, 208n38

Odyssey, 44
Ortelius, Abraham, 59
Orwell, George, 138
Ovesey, Lionel, 181

Pan Ku, 109
paraphier, 22, 24, 26, 27, 44, 49, 66, 78, 79, 186, 187
paraphrand, 22, 24, 26, 27, 29n15, 44, 49, 66, 78, 79, 186, 187
personality, 32, 36, 145, 149, 150, 155, 157, 180, 194
perspective, 158, 166
petites perceptions ("little perceptions"), 94
Petrarch, Francesco, 158
phrenology, 51, 52, 73
physiognomy, 73
physique/moral dualism, 83, 84
Plato, 41, 45, 47n13, 53
pneumatology, 72
Polsky, Andrew J., 185
positivism, 129, 130, 132, 139
pre-Socratic thought, 42, 43, 44
Prester John, 60
progress, 93, 111, 113, 118, 119, 120, 121, 122, 127, 128, 129, 130, 131, 132, 133n5, 135, 136, 139, 143, 144, 165. *See also* progressivism; secular humanistic progressivism
progressivism, 117, 124, 139. *See also* progress; secular humanistic progressivism
proto-Psychology, 71, 74, 75, 76, 95n8, 150, 193
psyche , 44, 45, 46, 47n10, 47n13, 52, 54, 72
psyche, ix, 4, 5, 6, 9, 10, 11, 12, 21, 23, 26, 29n12, 42, 51, 52, 56, 64, 71, 81, 85, 88, 90, 91, 92, 95, 117, 122, 133, 161, 162, 166, 169, 170, 177, 178, 180, 182n12, 182n14, 183, 184, 187, 193, 194, 199, 202, 206; as object of experimentation, 97, 98, 100, 102, 103; plasticity of, 198, 199; space and time, 14; spatialization of, 14, 51. *See also* mind
psychologia, 72
psychological idolatry, 182
psychological processes, ix, 4, 5, 6, 7, 11, 13, 16, 22, 24, 25, 28, 32, 34, 43, 44, 47n10, 52, 55, 56, 71, 72, 73, 75, 76, 84, 88, 97, 100, 104, 144, 145, 151, 158, 162, 169, 171, 173, 178, 179, 180, 181, 185, 186, 194, 195, 197, 199, 204, 205. *See also* interiority
psychological revolution, 183
psychology, vii, viii, ix, 5, 6, 11, 15, 20, 21, 27, 28n11, 32, 43, 46, 55, 56, 63, 71, 72, 73, 74, 75, 76n7, 77, 78, 79, 80, 83, 85, 97, 98, 99, 103, 104, 122, 131, 132, 150, 152, 165, 177, 180, 181, 190, 193, 194, 196n12, 197, 198, 199, 201, 202, 206; birth of, 89, 90, 91, 92, 93, 95, 169, 170, 171, 173, 174n3, 174n4;

folk, 33; physiological, 101, 102
psychophysical parallelism, 92, 103, 173
psychoscape, 10, 26, 32, 42, 145, 146, 163, 196n12, 206
Ptolemy, 59
Purgatory, 193, 195

qi (quasi-material life force), 54, 55
Quimby, Phineas Pankhurst, 178

Radding, Charles, 6, 56n4, 143, 152n3, 152n4, 152n6, 152n17, 153n20, 153n23, 153n24
Rappaport, Roy, 132
Ratherius of Verona, 155
rationalism, 16, 90, 94, 103, 135, 136
reactivity (perceptual), 36, 42, 202, 203, 208n3
realism, 66, 148
reframing, 19, 20, 163. *See also* meta-framing
Renaissance, 51, 63, 68n16, 71, 81, 82, 97, 114, 120, 127, 146, 152n11, 158, 164, 194
Richet, Charles, 138
Rieff, Philip, 184
Rogers, Carl, 181
Rolando, Luigi, 75
role, 185, 202, 206; diversification of, 161, 162, 164, 165, 166, 167, 167n2, 168n8; social, 82, 151, 155, 161. *See also* "I," "me," self
Roman Empire, 121, 143, 145, 147
Romans, 59
Romanticism, 16, 103, 196n9

Sahlin, Marshall, 17n11, 133
Saint-Simon, Count Henri de Rouvroy de, 128, 129, 130
Santorio, Santorio, 67
Scaliger, Joseph Justus, 114
Scholasticism, 148
scopic, 5, 11, 12, 21, 162; transition from cosmic, 24; worldview, 21. *See also* introscopic; macroscopic; microscopic
Scottish Enlightenment, 72, 91
secular humanistic progressivism, 117. *See also* progress; progressivism
self, 143, 144, 145, 149, 152, 158, 161, 162, 163, 164, 165, 166, 167, 167n6, 167n7, 168n8, 177, 179, 180, 181, 182, 183, 184, 185, 186, 187, 188, 189, 190, 191n4, 191n16, 194, 195, 199, 203, 205, 206, 207
self-authorization, 36, 97, 105n3, 144, 167
self-autonomy, 36, 80, 97, 98, 123, 144, 145, 156
self-consumption, 190
self-development, 15, 144, 186
self-esteem, 145, 181, 183
self-help, 139, 144, 177, 178, 179, 180, 181, 184, 185, 188, 189, 191n16; books on, 181, 185
self-identity, 16, 145, 181
self-improvement, 15, 144, 165, 183, 188, 191n17
self-narratization, 35, 36, 97, 144
self-ownership, 144, 189
self-reflexivity, 16, 36, 80, 97, 122, 133, 145, 156, 157, 161, 163, 164, 165, 166, 168n8
sensate experiences, 24
Service, E. R., 97, 133
sexual equality, 124
Shangri-la, 156
Sima Qian, 109
Simonides of Ceos, 50
Skinner, B. F., vii
Smiles, Samuel, 144
social (re)engineering, 118, 127, 189
social evolution, 119, 132, 139
soma , 43, 44, 45, 46, 85, 88
soul–mind, 42, 43, 45, 49, 52, 53, 81
Southern Continent, 60
space, imaginary, 3, 5, 14, 15, 21, 25, 26, 33, 34, 49, 63, 146, 149, 156, 195, 209, 211
spatialization: of history, 136; of mind, 10, 11, 34, 35, 41, 49, 51, 64, 81, 83, 92, 98, 156, 158, 163; of space, 15; of time, 3, 26, 27, 109, 111, 112, 114, 121. *See also* interiorization
spatio-visual experience, 4, 9, 10, 12, 23, 25, 31, 74
Spencer, Herbert, 93, 131
spirit possession, 98
spiritualist mediums, 98

Spurzheim, G., 51
St. Augustine, 3, 7n1, 46, 49, 51, 52, 53, 56n4, 56n5, 56n7, 56n10, 56n13, 76n3, 78, 111, 119, 124n2, 143, 146, 157, 158
St. Benedict, 143
St. Paul, 52
Staël, Madame de, 122
Stahl, Georg Ernst, 88
Steward, Julian, 133, 133n6
Stoics, 46
Sun Yat-sen, 124

Taiping Rebellion, 124
technology, 11, 15, 26, 100, 112, 123, 132, 136, 137, 138, 139, 190, 195. *See also* techno-science
techno-science, 11, 20, 33, 135; and rationalism, 136, 137. *See also* technology
tele-astronomical, 21. *See also* scopic; tele-scopic
tele-scopic, 21. *See also* scopic; tele-astronomical
Terman, L. M., 180
therapeutic theism, 178, 180
therapeutic turn, 178, 184
therapeutization, of daily life, 5
therapy-centric society, 179, 180
Thomistic interpretation, 49, 53, 81
Thomsen, Christian Jürgensen, 121
Thucydides, 110
time, linearization of, 12, 14, 22, 35, 61, 63, 110, 111, 112, 118, 127, 139, 144, 156. *See also* spatialization; time, spatialization of
time, spatialization of, 3, 26, 27, 109, 111, 112, 114, 121. *See also* time; linearization of
T-O maps, 60. *See also* wheel maps
triumphant suffering, 186
Tucker, James, 184
Turgot, Anne Robert Jacques, 121
Turner, Victor, 23, 29n18, 153n37
Twenge, Jean, 190
Tylor, Edward Burnett, 132

unconscious, 21, 32, 53, 56, 72, 88, 94, 152, 181. *See also* nonconscious,
unconscious inference, 101
utopia, 60, 117, 124, 129, 132, 135, 138, 166

Valla, Lorenzo, 114
Verne, Jules, 137
Vesalius, Andreas, 15, 66
victimization, 186
visions, 31, 61, 193, 195, 196n14
visual perception, 4, 9, 66, 92, 102
visuality, 4, 10, 11, 12, 14, 21, 23, 24, 31, 41, 51, 63, 64, 67, 74, 84, 97, 105n2, 158, 206
vitalism, 25, 81, 88
Vives, Juan Luis, 119, 145

Wang Yangming, 55
Ward, Steven C., 184
Watson, John B., vii
Watt, Henry J., 171
Weber, Ernst Heinrich, 73, 104, 194
Weintraub, Karl, 47n18, 68n14, 150, 153n32, 157, 159n1, 159n3, 159n4, 159n5, 159n9, 159n10, 159n11, 159n12, 159n18
Wells, H. G., 137, 138
Wertheim, Margaret, 68n7, 68n11, 68n14, 68n18, 173, 195, 196n1, 196n5, 196n10, 196n13
wheel maps, 60. *See also* T-O maps
White, Leslie, 132
Whytt, Robert, 72
Worsaae, Jens Jacob, 121
Wundt, Wilhelm, 94, 101, 104, 132, 169, 170, 171, 174n3
Würzburg School, 171, 172, 174n4

Yates, Frances A., 56n8, 56n9

About the Author

Brian J. McVeigh received his PhD in anthropology from Princeton University. A specialist in Japan and China, he lived and taught in Asia for seventeen years. He is now training in mental health counseling at the University at Albany, SUNY. He is currently writing *The "Other Psychology" of Julian Jaynes: Cultural–Historical Approaches to Cognitive Adaptation and Worldwide Patterns of Religiosity* and *When God Talked among Us: Biblical Evidence of Bicameral Mentality.* His others books include:

The History of Japanese Psychology: Global Perspectives, 1875–1950
How Religion Evolved: Explaining the Living Dead, Talking Idols, and Mesmerizing Monuments
The Propertied Self: A Psychology of Economic History
Interpreting Japan: Approaches and Applications for the Classroom
The State Bearing Gifts: Deception and Disaffection in Japanese Higher Education
Nationalisms of Japan: Managing and Mystifying Identity
Japanese Higher Education as Myth
Wearing Ideology: State, Schooling, and Self-Presentation in Japan
The Nature of the Japanese State: Rationality and Rituality
Life in a Japanese Women's College: Learning to Be Ladylike
Spirits, Selves, and Subjectivity in a Japanese New Religion: The Cultural Psychology of Belief in Sūkyō Mahikari

www.ingramcontent.com/pod-product-compliance
Ingram Content Group UK Ltd.
Pitfield, Milton Keynes, MK11 3LW, UK
UKHW041347230926
475560UK00013B/70

9 781498 520287